STRUCTURAL RELIABILISM

TRENDS IN LOGIC
Studia Logica Library

VOLUME 16

SCOPE OF THE SERIES

Trends in Logic is a bookseries covering essentially the same area as the journal *Studia Logica* – that is, contemporary formal logic and its applications and relations to other disciplines. These include artificial intelligence, informatics, cognitive science, philosophy of science, and the philosophy of language. However, this list is not exhaustive, moreover, the range of applications, comparisons and sources of inspiration is open and evolves over time.

Volume Editor
Ryszard Wójcicki

The titles published in this series are listed at the end of this volume.

PAWEŁ KAWALEC
The Catholic University of Lublin, Poland

STRUCTURAL RELIABILISM

Inductive Logic as a Theory of Justification

SPRINGER-SCIENCE+BUSINESS MEDIA, B.V.

A C.I.P. Catalogue record for this book is available from the Library of Congress.

ISBN 978-94-010-3955-0 ISBN 978-94-010-0259-2
DOI 10.1007/978-94-010-0259-2

Printed on acid-free paper

To my Mother Krystyna Kawalec

Contents

Preface

The human quest for knowledge is persistently accompanied by the problem of justification. Notwithstanding the prodigious enrichment of the body of human knowledge, we fall short of concluding the philosophical debate on the nature of knowledge and justification. In particular, the emergence of a philosophical theory of justification which will afford us hope of bringing the prolonged debate to an end still seems unlikely. While more and more sophisticated epistemological theories of justification are put forward, they seem to fall within the broad category of *mentalistic* theories. By contrast, this book focuses on inductive logic as a genuine alternative to the aforementioned theories of justification.

As the title indicates, I am preoccupied here with an epistemological analysis of a *theory* of justification. To provide an epistemological *theory* of justification is tantamount to providing an informative general statement of the application conditions of justification. In particular, I examine a kind of *reliabilist* theory of justification. Reliabilist theories propound the view that epistemic justification supervenes on non-epistemic properties of justification-bearers. The *structural* reliabilist theory claims in addition that justification supervenes on syntactic and semantic properties of sentences as justification-bearers.

According to R. Carnap deductive logic and inductive logic constitute *the* theory of justification of unified science:

> But still some question of justification must be raised; that is the task of the philosopher. But I would like to give the question a form different from the customary one. If we search for a justification, the first thing we have to make clear to ourselves is this: to which particular point or points in the system of inductive procedure should we direct our scrutiny? ... Many believe that the essence of

> inductive reasoning is to make inferences of a special kind. From this point of view, the demand of justification would naturally be directed at any alleged principles of inductive inference. I would not say that it is wrong to regard inference making as the aim. But from my point of view it seems preferable to take as the essential point in inductive reasoning the determination of probability values. ... In my view inductive logic is the theory of this particular concept of probability (Carnap 1968a, 258-59).

The main problem of the book concerns the kind of theory of justification propounded in the form of the systems of inductive logic. I seek to contrive a general statement of the application conditions of justification on the grounds of the systems of inductive logic developed by Carnap and the members of the Finnish School, most notably by J. Hintikka.

I claim that the systems of inductive logic provide the grounds for the general statement of a necessary condition of justification. This condition refers to syntactic and semantic properties, or on the whole to the *structural* properties of sentences which are justification-bearers. Since in the epistemological theory represented in the systems of inductive logic justification supervenes on structural properties in a way consonant with reliabilism I find it appropriate to refer to this theory as a *structural reliabilist* theory of justification.

What from the perspective of the mainstream epistemology might appear as disqualifying my claim that the systems of inductive logic propound, as a matter of fact, an epistemological theory of justification is that sentences are not taken to constitute the proper kind of justification-bearers. For it is primarily beliefs, or other mental states which have propositional contents, that are justification-bearers. I refer to theories of justification which embrace the latter view as *mentalistic* theories. On the other hand, however, the traditional epistemological standpoint—still persistent in the Polish philosophical tradition which draws upon the heritage of the Lvov-Warsaw School—holds that it is propositions which require epistemological scrutiny in the first instance. I refer to the group of theories of justification which draw upon syntactic and semantic properties of justification-bearers as *minimalist* theories.

Throughout my work I demonstrate that, contrary to what is commonly assumed, substantive claims concerning the conditions of justification can be advanced on the grounds of the *minimalist* theory of justification brought out by Carnap and the Finnish School in the form of the systems of inductive logic, i.e. on the grounds of *structural reliabilism*. Therefore I seek to demonstrate that structural reliabilism is a coherent epistemological position. The key objection to structural reliabilism leveled on the grounds of mentalistic theories of justification is the Goodman paradox. In order to respond to this objection I discuss in detail how to resolve this paradox on the grounds of structural reliabilism. I also delineate the essential points of advantage that structural reliabilism enjoys over its mainstream competitors.

Despite the severe criticisms of the whole program of inductive logic, no one has yet undertaken the task of setting forth an appropriate epistemological analysis of the Carnap-Hintikka style systems of inductive logic as *constituting*, in fact, an *epistemological* position, although the view that "the relation ... of confirmation is ... transparently epistemological" (Creath 1992, 146) has recently been widely acknowledged among Carnap scholars (Friedman 1992, 17). Inductive logic is by and large perceived as an effort to elaborate on Hempel's model of explanation in order to explicate the concept of hypothesis confirmation.[1] Alternatively, inductive logic is perceived as an attempt to develop a criterion of meaning which will alleviate the restrictive criterion of verification, which was initially advocated by logical empiricists. The concept of confirmation, contrary to the concept of verification, is a gradual one, and therefore more suitable as a criterion of meaning.[2] Finally, there is also the view advocated by W. C. Salmon, who claims that Carnap gave rise to the whole program of inductive logic in order to provide a solid foundation for statistical inferences (Salmon 1967b; 1990a). What all the three approaches to the program of inductive logic have in common is a deficiency in bringing out the epistemological issues prevailing in this program. In effect, Carnap's epistemological position has often been outrageously oversimplified and misconceived,[3] and moreover, the mentalistic character of the mainstream theories of justification enjoys the status of an unquestioned dogma,[4] even in the face of their apparent failure to come to grips with the pressing problem of justification.

In general, I contrive the framework for the structural reliabilist theory of justification by critical assessment of meta-level properties underlying the mentalistic version of reliabilism such as being a *substantive, nomic* and *quietistic* theory of justification. Only then do I proceed to project the general statement of a necessary condition of justification on the grounds of structural reliabilism. This condition is cast in structural terms, and therefore turns on the technical details of the construction of the systems of inductive logic developed by Carnap and the Finnish School.

The first chapter, *The Systems of Inductive Logic*, delineates the details of the construction of the systems of inductive logic which are germane to the later discussion. In the first section I discuss conceptual advances in philosophy and probability theory pertinent to the development of the *quantitative* systems of inductive logic by Carnap and the Finnish School, and which display the traditional philosophical concern for developing inductive logic, in general, and quantitative inductive logic, in particular. In the remaining sections I proceed then to present the development of the

Carnap-Hintikka style systems of inductive logic. I attempt to delineate between Carnap's versions of the systems of inductive logic and the versions thereof accomplished by Hintikka and his co-workers, on the one hand, and also to delineate between the initial formulations of the systems of inductive logic and the recent Bayesian reformulation thereof, on the other hand. The distinction between the initial formulation of the systems and their Bayesian reformulation seems especially important for the epistemological interpretation of the systems I develop in this book.

In the second chapter, *Inductive Logic as a Structural Reliabilist Theory of Justification,* I address the meta-level characteristics of theories of justification. I define two kinds of concepts of epistemic justification:[5] *mentalistic* and *minimalist*. The necessary condition for a mentalistic theory of justification stipulated in the first section is that in this theory justification is attributed to doxastic attitudes,[6] and primarily to beliefs. In the second section I then proceed to demonstrate that a mentalistic theory of justification, in particular A. I. Goldman's process reliabilism, is based on conflicting meta-level assumptions, which render this kind of theory incoherent. In the third section I define the minimalist theory of justification. The necessary condition for a minimalist theory of justification is that in this theory justification is attributed to sentences, in particular sentences accommodated within the body of scientific knowledge. In this section I also set up the structural reliabilist framework of the necessary condition of justification in the form of the recursive clause.

The primary task I undertake in the fourth and fifth section is a detailed analysis of how the framework of the structural reliabilist position elucidated thus far is realized and argued for in the systems of inductive logic. I refer to the most severe criticisms of inductive logic in order to point out that they presuppose that the Carnap-Hintikka style systems of inductive logic represent an empiricist epistemological position. I argue then that this epistemological interpretation, in particular a foundationalist or coherentist view, is highly inadequate. I then proceed to provide arguments to substantiate the view that the epistemological position constructed by means of the systems of inductive logic is structural reliabilism. The essential feature of this position consists in attributing to logic a special epistemic status, which conflicts with the empiricist view on the nature of logic. It is the logical structure of the language in which a given sentence originated that determines the justificatory status of this sentence. By means of extra-logical assumptions Carnap and Hintikka are able to establish a numerical measure of the justificatory status of the sentences embedded in a given language. They also provide clues on how to proceed from a minimalist theory of justification, to a mentalistic theory of justification which will be pertinent to the tripartite definition of knowledge.[7]

Finally, in the third chapter, *Structural Reliabilism and the Problems of Bayesian Epistemology,* I proceed to analyze whether and how the structural reliabilist theory is able to confront epistemological problems. In the first section I repudiate the main epistemological criticisms against structural reliabilism which are based on a confusion between mentalistic and minimalist theories of justification. This confusion turns on the mistaken view of Carnap as an heir to the empiricist tradition. In the second section I focus on the generic problems of structural reliabilism, namely, the problems which any Bayesian epistemology must face. In the series of arguments I demonstrate that structural reliabilism constitutes a coherent epistemological alternative, which is relevant to the central task of epistemology, to wit, the definition of the concept of knowledge. I also examine whether it is able to meet the objections raised against mentalistic versions of reliabilism. The most important argument against the latter claim is recognized as the Goodman paradox, discussed in the third section, which grounded in epistemology the wide-spread view that the systems of inductive logic are irrelevant for epistemology. Drawing, in part, upon B. Skyrms's argument, I demonstrate how the Goodman paradox is resolved in the systems of inductive logic. I also discuss how structural reliabilism responds to the longstanding problems of confirmation. In the fourth section I discuss how structural reliabilism addresses one of the most severe difficulties of Bayesian epistemology, i.e. the prolonged debate on whether that which determines a prior distribution of probabilities is something characteristically objective or subjective.

Since the year 2000, when the initial version of this manuscript was completed, we observe a steadily growing research interest in logical probability,[8] and I can only hope that this book will further stimulate it.

Notes

1 Cf. (Lehrer 1970); (Kuipers 1997). W. C. Salmon claims that Carnap's concept of the degree of confirmation is at odds with Hempel's model of explanation, because of the problem with zero-confirmation of generalizations; cf. (Salmon 1965b, 140; 1990a, 105-106).

2 Cf. (Rynin 1947); (Walker 1947, 99); (Hempel 1950/1996; 1950/1959, 111; 1975, xix); (Carnap 1956/1996); (Wójcicki 1966/1977); (Misak 1996, 73-74); (Kuipers 1998, 532). See (Hesse 1969, 91-92) for an argument that the concept of confirmation was never intended by Carnap to fulfill this role.

3 For a survey of epistemological interpretations of Carnap's works see (Moulines 1991, 264).

4 Remarkably, all theories of the epistemic theories of justification discussed by R. Chisholm in (Chisholm 1982, 29-32) are—what I propose to call—*mentalistic* theories of this concept; see also (Mann et al. 1992, 256); (Plantinga 1993; 1995a; 1995b).

5 For a demarcation between epistemic and non-epistemic concepts of justification see e.g. (Foley 1998, 158).

6 According to R. Chisholm, there are three relevant types of doxastic attitudes an epistemic agent can assume: to believe, to disbelieve, and to withhold belief; cf. (Chisholm 1966/1977, 6).

7 Cf. (Plantinga 1993, 6). For a historical account see (Wójcicki 1997).

8 Cf. (Franklin 2001); (Bartha and Johns 2001).

Acknowledgements

I am indebted to many people for their comments and discussions on various parts of the book as well as on the papers that preceded it: Raymond Dacey, Roberto Festa, Józef Herbut, Risto Hilpinen, Adam Grobler, Elżbieta Kałuszyńska, I. Niiniluoto, Wlodek Rabinowicz, Wesley C. Salmon, and Sandy L. Zabell. I want also to thank my colleagues from the Department of Methodology at the Catholic University of Lublin for stimulating discussions.

I should especially like to thank Andrzej Bronk, Leon Koj, Theo A. F. Kuipers, and Ryszard Wójcicki, who provided detailed comments on earlier drafts of the book.

I want also to express my warm gratitude to my wife Anna for her enduring support.

I would like to thank the editors: Charles Erkelens, Rudolf Rijgersberg, Jolanda Voogd and Tamara Welschot of Kluwer Academic Publishers for their invaluable assistance, and also to the anonymous reader for the helpful suggestions and constructive criticisms.

Chapter 1

THE SYSTEMS OF INDUCTIVE LOGIC

R. Carnap succinctly characterizes his meta-epistemological views as follows:

> the theory of knowledge, which is after all nothing but applied logic, can no more dispense with symbolic logic than physics can dispense with mathematics (Carnap 1930/1959, 137).

According to this view, "the theory of knowledge" as an "applied logic" turns on the construction of the systems of purely formal logic. The systems of purely *formal* logic do not constitute the theory of knowledge, but they represent the theory of the justificatory structure of sentences or propositions, which constitutes a necessary condition for the theory of knowledge. In the view of obtaining a comprehensive representation of the justificatory structure Carnap set himself the task of developing not only the systems of deductive logic, but also of constructing the systems of *inductive* logic.[1] The systems of inductive logic that Carnap developed in the early '50s were subsequently generalized by J. Hintikka and his co-workers.

In this chapter I will focus on the technical details of the Carnap-Hintikka style systems of inductive logic. This will be done essentially in three steps. In section 1 I introduce the philosophical background of the discussions preceding Carnap's and Hintikka's constructions of the systems of inductive logic. There seem to be two major components to this background worth highlighting: advances in logic, and developments in probability theory. Section 2 tracks the successive steps in the evolution of Carnap's and Hintikka's versions of the systems of inductive logic. And finally in section 3 the subsequent developments of the systems are delineated.

The detailed discussion of the theory of justification represented by the systems of inductive logic will be presented in chapters 2 and 3.

1. The Main Ideas Preceding the Construction of the Systems of Inductive Logic

It is beyond the scope of the present book to provide a comprehensive historical survey of the advances in inductive logic. I will limit the presentation to what seems directly relevant to the study of the Carnap-Hintikka style systems of inductive logic. Of course, while any boundary between what is considered as relevant and not will be to some extent arbitrary, it would perhaps be most convenient to limit the presentation to those conceptual developments which were necessary for the versions of inductive logic under the study.[2]

Notwithstanding the fact that credit for being first to provide a theory of inductive inference and a prevailing definition thereof belongs to Aristotle, it is still debatable what was precisely the concept of induction embraced by him.[3] Since then there have been basically two sorts of conceptual development pertaining to the epistemological position espoused in the systems of inductive logic studied in this book. The first is the line of radically changing views on what might constitute the necessary and sufficient conditions for inferential justification. Interestingly, R. Descartes emphasis on the indispensable role of an epistemic agent's internal states in establishing the justificatory status—an idea so foreign to Aristotle's conception of demonstrative *epistéme*—proved extremely influential. This idea was inherited by J. Locke, D. Hume and J. S. Mill, and then by 19th century German philosophy and logic.[4] Admittedly, it substantiated mentalistic, as contrasted here with minimalist, theories of justification.

To some extent, as will become more perspicuous in the chapters to follow, this "deviation" from the traditional, or logical for that matter, framework for resolving epistemological questions has been remedied by the second conceptual setting, i.e. the gradual development of probability theory. I. Hacking's thorough study (Hacking 1975) clearly depicts the fact that probability theory has always been motivated by vital epistemological problems. For on the one hand, so it seems, there are objective chance mechanisms operating out there in the world which manifest themselves, for instance, in tossing a coin with unknown bias, while on the other, there are gradual belief-states or confidence-states, which are sometimes due to our lack of relevant information on what has happened out there in the world. Hacking recognizes as one of the key points in the history of the probability theory the combination of the two, and the application of objective chance mechanisms to map the subjective gradual belief-states. Hacking writes:

> Probability has two aspects. It is combined with the degree of belief warranted by evidence, and it is connected with the tendency, displayed by some chance devices, to

> produce stable relative frequencies. Neither of these aspects was self-consciously and deliberately apprehended by any substantial body of thinkers before the time of Pascal (Hacking 1975, 1).

As the contemporary disputes between various interpretations of probability, and the disputes between classical and Bayesian statisticians point out the epistemological problem of how to combine the "two aspects" seems far from being resolved. Moreover, there are still parties of the dispute that are at pains to emphasize exclusively one aspect of the concept of probability: personalists—like C. Howson, B. Skyrms and R. C. Jeffrey—the subjective aspect, while frequentists—like W. C. Salmon—the objective aspect. Carnap was the first to legitimize the dichotomous nature of the concept of probability (Carnap 1945; 1945/1970). On the one hand there is a *logical* and analytical concept of probability, and, on the other hand, there is an *empirical* and synthetic concept of probability. The former belongs to the province of formal logic, especially semantics, while the latter belongs to the province of empirical science. Carnap's idea how to merge the two is elicited in the last section of this book.

According to Hacking, there seems to be a close conceptual affinity between Carnap's logical concept of probability and G. W. Leibniz's concept thereof (Hacking 1967). Hacking claims that seven elementary features of Carnap's program of inductive logic were already present in Leibniz:

> First, there is such a thing as non-deductive evidence... Second, "being a good reason for" is a relation between propositions. Third, this relation is to be characterized by a relation between sentences in a suitably formalized language. Fourth, there is an ordering of reasons from good to bad... Fifth, this measure is ... independent of anyone's opinion... Sixth, this measure is global—it applies to any pair of propositions... Moreover, this global relation is a "formal" one (Hacking 1975, 134-35).

Leibniz, in contrast to such contemporaries as B. Pascal, A. Arnauld and P. Nicole, in his studies on probability was not motivated by practical problems of gambling (Hacking 1975, 85). He was concerned, rather, with classes of statements such as law statements, which do not express absolute, but only contingent truths, and which are always conditional on evidence. Hence, "Leibniz took numerical probability as a primarily epistemic notion" (Hacking 1975, 89), i.e. as a measure of the degrees of uncertainty rather than a measure of physical chances in a game set-up. In his study on conditionals (*De conditionibus*, 1665) Leibniz establishes the idea that probability theory with respect to contingent truths might serve the role the theory of deduction serves with respect to necessary truths. The theory of absolute truths was well established; what was still lacking, however, was a theory of contingent truths, i.e. a theory of *probabilitas*, to use Leibniz's own word. The central point consisted in assigning numbers to various statements according to the degrees of probability they express: 1 being an absolute statement, 0 an impossible statement, and a rational number ranging between 0 and 1 representing a contingent statement.[5] Suppose we have a conditional statement $p \rightarrow q$, where p is equivalent to a disjunction of alternatives: $p_1 \vee p_2 \vee \ldots \vee p_n$. If it is the case that each of the disjuncts $p_1, p_2, \ldots, p_n$ implies q or implies a sufficient condition for q, then, according to Leibniz, $p \rightarrow q$ is assigned the value of 1. If, however, it is the case that none of the disjuncts implies q nor q's sufficient conditions, then $p \rightarrow q$ is assigned the value of 0. Finally, it is possible that some of the disjuncts imply q or its sufficient condition, while the rest implies $\neg$q, or its sufficient condition. In such a case $p \rightarrow q$ should be assigned, according to Leibniz, a rational number between 0 and 1. By allowing thus for the possibility of measuring uncertainty he

> was the first to believe in a formal logic of induction that is to be advanced by constructing artificial languages of unified science... Leibniz's injunction "calculate!" is not the cry of some alien rationalism, but a cry for inductive logic (Hacking 1967, 398).

It was a common belief in the 17th century that the uncovering of the exhaustive set of elementary ideas coupled with the universal grammar prescribing how to combine these into more complex ideas would open up a way to reveal all the secrets of nature (Hacking 1975, 140). In much the same vein Leibniz expected that once we have the *characteristica universalis,* i.e. the exhaustive set of simple ideas necessary and sufficient to generate all complex ideas, we will be able, by means of art of combinations, to generate descriptions of all possible worlds. Since all these possible worlds have some propensity to exist, it seems most reasonable for Leibniz[6] that they all should be assigned equal probabilities. Accordingly, Leibniz in 1678 defines probability by means of what was then called *the principle of indifference,* i.e. as "the ratio of favorable cases to the total number of equally possible cases" (Hacking 1975, 122). Thus, the idea of a probability measure function over a set of all possible worlds, and a set of all sentences, so fundamental for Carnap, was clearly anticipated by Leibniz.

In *Meaning and Necessity* (Carnap 1956) Carnap admits that his semantic concept of "state-description" is due to L. Wittgenstein's idea of possible-worlds, and hence indirectly to Leibniz. Hintikka, on the other hand, contests that Carnap was only close to develop the possible-world semantics, but failed "to interpret his models as genuine ... alternatives to our actual world" (Hintikka 1973/1975, 220).[7]

If one accepts Leibniz's idea of assigning a probability measure to every sentence and every pair of sentences, then, on the standard account, one therewith warrants T. Bayes's[8] theorem a wide area of applications:

$$P(h \mid e) = P(h) \cdot P(e \mid h) / P(e), \qquad \text{(Bayes's Theorem)}$$

which states that the probability of a hypothesis *h* conditional on the evidence *e* equals the *prior* probability P(*h*) of *h* times the *likelihood*, or the *inverse probability*, $P(e \mid h)$ of *e* on *h* divided by the *total probability* P(*e*) of *e*. The latter, called also the probability of *total evidence*, summarizes the relevant evidence we obtain after performing an experiment, or a series of experiments. A prior probability of a sentence, or its *unconditional* probability, is the probability assigned to a given sentence prior to performing an experiment. Prior probability is often interpreted as a summary of the background knowledge. According to this view the prior probability I assign to the sentence "This wheel of fortune is loaded" summarizes my background knowledge on similar wheels of fortune which I possess prior to spinning the wheel.

Bayes's theorem forces one to assign prior probabilities to any sentence, i.e. to measure the probability of a given sentence prior to any experiments undertaken in order to verify it. Moreover, it forces one to adopt a probability measure defined for *conditional* and *inverse conditional* relations between every two sentences. Bayes's theorem seems to accord well with Leibniz's idea of a probability measure defined for every sentence, i.e. for every simple and every complex sentence. The problem for Leibniz, however, was how to assign numbers to contingent statements, and we have seen that he made it dependent on the success of the *characteristica universalis*.

S. Zabell, in investigating the precise date when Bayes first formulated the theorem, comes to the conclusion that Bayes thought of it as a response to Hume's problem of induction posed in 1748 in his *An Enquiry Concerning Human Understanding*.[9] It is Laplace, however, who takes credit for applying probability theory in general, and Bayes's theorem in particular, to inductive inference. He employed it in calculating the probability of the next sun-rise given all previous sun-rises. In this calculation Laplace used the notorious rule of succession:

$$p(h_i \mid e_m) = (m+1) / (m+2); \qquad \text{(The Rule of Succession)}$$

which determines the probability of an event *i* given it is reported in *e* to have successively occurred *m* times. Zabell claims that Laplace's calculation of the probability of the next sun-rise, so often then to become a subject of ridicule, was motivated basically by Hume's assertion in his *A Treatise of Human Nature* (Hume 1739/1967) to the effect that the next rising of the sun exceeds all probability, although is not demonstratively certain (Zabell 1989, 293). Hume's assertion triggered off the long debate on whether inductive inference is amenable to probability theory, and if so, whether it still can be approached in quantitative terms (Zabell 1989, 293-304).

It is only at the end of the 19th century and the beginning of the 20th that we observe a growing assimilation of Laplace's probabilistic approach to inductive inference. The general probabilistic framework in approaching inductive inference differentiates respective to interpretations of the concept of probability. First, there is the classical concept of probability introduced by Laplace, and often ridiculed afterwards, which states that:

> if there is no reason to expect any of the elementary events E to be more probable than others (the *principle of indifference*), then all elementary events are equally probable.[10]

Suppose we have a regular die with numbers from 1 to 6 on its sides. The classical definition of probability requires that each of the elementary outcomes of one throw of the die be assigned equal probability, i.e. 1/6.[11]

Second, there is the frequency concept of probability developed by J. Venn (Venn 1888/1972), R. von Mises (Mises 1957), and H. Reichenbach (Reichenbach 1935/1949).[12] According to Mises, probability is, roughly speaking, the limit of the frequency of successes relative to all performed trials. Suppose that in throwing the above described die 60000 times the face with number 1 on it turned up 10020 times. The frequency definition of probability requires that the probability of throwing 1 on this die be equal to the limit of the ratio 10020/60000 (with the number of throws tending to infinity), i.e. 1/6 again.

The classical and frequency concepts of probability bear close affinity to the original problems that gave rise to the probability theory. Leibniz, as noted earlier, established much his own approach, because he was more concerned with probabilities of sentences (or propositions) rather than events.

Hacking observes that Carnap overtook from his predecessors J. Keynes and H. Jeffreys the program of inductive logic, to a large extent anticipated by Leibniz.[13] Leibniz's idea of theory of probability as the "new kind of logic" was dormant until the 1920s, when Keynes and Jeffreys refurbished it.

Keynes[14] is credited with the logical approach to probability. Recall the die example. There are six sentences S_i describing the elementary outcomes of throwing the die, each of the form "The face with number i turns up", where i is one of the integers from 1 to 6. The disjunction

$S_1 \vee S_2 \vee S_3 \vee S_4 \vee S_5 \vee S_6$ is assigned the value of 1, its negation $\neg S_1 \wedge \neg S_2 \wedge \neg S_3 \wedge \neg S_4 \wedge \neg S_5 \wedge \neg S_6$ the value of 0, and each of the S_i sentences has a probability equal to 1/6. Keynes, however, was highly skeptical about Leibniz's idea of defining probability measure over the set of all sentences.[15] He thought it impossible in many of cases, and proposed the comparative rather than quantitative concept of probability in these cases. According to Keynes then even if it does not make sense to ascribe a numerical probability to a given sentence, it still makes sense to compare its probability with the probability of some other sentence.

The difficulties of each of the interpretations of probability led A. N. Kolmogorov to formulate an elementary set of axioms, i.e. an implicit definition, for the probability concept, which purportedly was independent of any of the interpretations of the probability concept.[16] Let $\mathcal{B}$ be a σ-field of sets of subsets of Ω. The standard axiomatization of probability[17] requires that for any $A, B \in \mathcal{B}$ the following hold:

$$0 \leq p(A) \leq 1; \qquad \text{(Axiom 1)}$$

$$p(\Omega) = 1; \qquad \text{(Axiom 2)}$$

$$p(A \cup B) = p(A) + p(B), \text{ if } A \cap B = \varnothing. \qquad \text{(Axiom 3)}$$

The last axiom, σ-additivity is sometimes admitted as an "expedient":

$$p(\cup_i A_i) = \sum_i p(A_i), \text{ if } (A_i \cap A_j) = \varnothing \text{ for any } i \neq j. \qquad \text{(Axiom 4)}$$

It was not only the development of mathematical logic, starting from late the 19th century,[18] that enabled Carnap to raise high expectations with regard to the accomplishment of Leibniz's idea of probability measure on the set of all sentences. The key event between Keynes's work and Carnap's is the development of formal semantics, starting basically with A. Tarski's definition of truth.[19]

2. Carnap's Systems of Inductive Logic

In his intellectual autobiography Carnap testifies that he became preoccupied with the problem of probability in 1941 (Carnap 1963a, 63). And this problem remained on the framework till his death in 1970.[20] In the late '40s he was still occupied with the semantical issues—what resulted in the book *Meaning and Necessity* (Carnap 1956a) which set out the agenda for his subsequent work on inductive logic—and largely under the influence of Leibniz, on the one hand, and Wittgenstein, on the other. While Carnap was attracted by Leibniz's idea of defining necessary truth in terms of possible-worlds, he embraced from Wittgenstein the idea of tautology as a complex sentence, which retains the truth value 1 on all possible substitutions of truth values to its constituent-sentences—an idea which led him to give birth to the concept of state-description.[21]

The first systematic attempt by Carnap to formulate a system of inductive logic is in his *Logical Foundations of Probability* (Carnap 1950). In his search for the foundations of non-demonstrative inferences Carnap identifies two legitimate concepts, so called "explicanda", of probability. One concept is an empirical concept of probability defined in terms of the limit of relative frequency. This kind of concept was identified by Carnap with Reichenbach's concept of probability. The empirical concept of probability was inappropriate for the kind of foundational research Carnap set forth for himself. Thus, he was more concerned with the other, i.e. logical, concept of probability. The appropriate explicandum leading to the definition of the logical concept of probability was identified by Carnap as "the degree of confirmation",[22] i.e. the degree of support an evidence statement e lends to a hypothesis h. The logical concept of probability then is the meta-language concept, which makes it different from the object-language empirical concept of probability (Carnap 1963a, 73). Some authors have identified the concept of the degree of confirmation as the measure of partial implication of e from h.[23] C. G. Hempel, for instance, succinctly characterizes Carnap's concept of confirmation thus: it "parallels and generalizes the formal conception of logical implication as a relation between sentences (rather than, say, between propositions); indeed, logical implication may then be regarded as a subrelation of confirmation" (Hempel 1967, 185). For it might appear natural to assign measure 1 to the confirmation of h by e if e logically implies h, and 0 in the case when e is logically implies $\neg h$, and some measure in-between 0 and 1 for all cases in which h and e are logically compatible, but no relation of logical implication obtains between e and h.

Let me now introduce the conceptual apparatus that Carnap employed in developing his subsequent versions of inductive logic.[24] In *Logical Foundations of Probability* Carnap elaborates on the system of inductive logic within a part of first order logic with identity. He takes into account only one- and two-argument predicates. Moreover, he does not consider a single language, but a family of languages, which differ with respect to the number of individual constants in the given language. Thus a "language system" is a tuple $<L, N>$, where L signifies a given subset of first order logic, and N the number of individual constants (e.g. $a_1, a_2, ..., a_N$) in L.

The two important differences between language systems in Carnap's account (Carnap 1950, 58) are the following: first, with respect to L the most interesting and actually worked out in detail case in *Logical Foundations of Probability* is the subset of first order logic without identity and only with one-argument predicates; and second, with respect to N the crucial division is between finite and infinite values thereof, which is particularly important as regards the degree of confirmation of generalizations, which in the latter case all have the value of 0. The fact that L is relative to N within a given language system has important consequences. It is not only that individual variables have different meanings in different language systems, but also that the universal quantifier "$\forall$" changes its meaning in different language systems. Suppose we have the expression "$\forall P(x)$". In any language system with finite N the expression is equivalent to the following conjunction: "$Pa_1 \wedge Pa_2 \wedge \ldots \wedge Pa_N$". And since it is impossible to express the conjunction in the case of the infinite N, Carnap seems to have been motivated to allow a number of language systems, which might be ordered sequentially with respect to N:

$$N = 1 \subset N = 2 \subset \ldots \subset N = \infty,$$

where all finite cases approximate more and more closely the infinite language system.

Moreover, Carnap decided to put further extra-logical constraints on language systems. One of them concerns the individuals of a given domain:

PROPOSITION 1.1 *Extra-logical constraint 1*

the individual constants of a given language system exhaust all individuals in "the domain of discourse" of the language system in question (Carnap 1950, 58).

The other constraints the number of predicates:

PROPOSITION 1.2 *Extra-logical constraint 2*

the number of predicates be finite (Carnap 1950, 59; 65).

As has been already emphasized, the language systems are not exclusively syntactical constructions. Each of them is in fact "a skeleton of a semantical system" (Carnap 1950, 59), because individual constants and predicates are intended to have an interpretation, but are not provided with it in advance of a given application.[25] Assuming then that they will be provided interpretation in a given application, Carnap continues to state the "rules of truth" for the formulas of a given language system. Then he comments thus:

> The semantical concept of truth ... [employed in the rules of truth] is such that the statement "the sentence i is true in L" in the metalanguage conveys the same factual information as the sentence i itself, which belongs to the object language (Carnap 1950, 69).

In view of the discussion of epistemological assumptions in Carnap's system it is important to note that the disquotational schema, which displays the implications of the semantical concept of truth, accounts for the functional role of the concept of truth, as well as the property of being "semantical", within the system of inductive logic in *Logical Foundations of Probability*. Remarkably, the sentences "$c(h \mid e) = r$" stating the degree of confirmation are, according to Carnap, meta-linguistic expressions themselves and only the reliance on the semantical definition of truth could grant that the contents of the factual argument-sentences (*h* and *e* in this case) remain unchanged in the process of the assessment of the respective degree of confirmation. There are two reasons why this point should be emphasized. First, it perhaps most perspicuously illustrates how, on Carnap's view, the standards of epistemic evaluation could supervene on the structure of given sentences and even embrace the latter. Second, this point substantiates the objection to some reformulations of Carnap's and Hintikka's systems of inductive logic. For it seems that the reformulations which result in the fact that the sentences stating the degree of confirmation turn out to belong to an object-language are not entirely faithful to the core epistemological project.

In constructing a system of inductive logic Carnap thinks of it as very much akin to a *quantitative* theory. We already know from Leibniz's analysis how to proceed in order to limit the range of numerical values ascribed to sentences within the interval [0, 1]. Carnap decided to apply this analysis within his own system, i.e. to ascribe 0 to all sentences which hold in none of possible worlds, and 1 to those which hold in all possible worlds. In order to achieve this Carnap needed first, a precise definition of a possible world, and, second, a canonical form for each of the sentences in the language system, which will univocally determine the set of possible worlds in which a given sentence holds. We thus proceed to analysis of two fundamental semantical concepts in Carnap's system, i.e. the concept of *state-description* and the concept of the *range* of a given sentence.

A state-description is a description of one of the possible configurations of individuals and their properties and relations in a domain of discourse *U* of *L*. Carnap then contends that the concept of state-description "must be taken as relative to a language system", more precisely, to "a logically possible state of a domain of individuals of L with respect to all attributes (properties and relations) designated by the [predicates] of L" (Carnap 1950, 71). Let LS^{π} be a language system with π primitive predicates $P_1(x), \ldots, P_{\pi}(x)$. These might be combined as follows to form an important complex predicate:

DEFINITION 1.1 *Q-predicate of* LS^{π}

$Q_i(x)$ *is a* Q-predicate *iff* $Q_i(x)$ *is a conjunction all of the primitive predicates* $P_1(x), \ldots, P_{\pi}(x)$, *each of which occurs once, negated or not.*

As this definition has it, any two Q-predicates may differ only as regards distribution of negation among the primitive predicates.

By appropriate application of Q-predicates to the domain of a language system LS_N (where N is the number of individuals in this domain, and $a_1, \ldots, a_N$ are constants representing those individuals) we obtain the following definition of *state-descriptions*:

DEFINITION 1.2 *State-description*

A state-description S_i *is a conjunction of* N *Q-predicates of* LS_N *such that for any* $i \in N$ $Q(a_i)$ occurs in the conjunction.

As a matter of illustration, let me invoke Carnap's own example here. Suppose we have a language system $L\,S_3^1$ with one predicate "*P*" and three individual constants "*a*", "*b*", and "*c*". One of the state-descriptions within this language system is $Pa \wedge Pb \wedge Pc$.

One of the serious contributors to the advancement of Carnap's systems of inductive logic—J. G. Kemeny—recognizes the crucial role of the concept of state-description for the construction of the systems:

> The entire method of Carnap's inductive logic can be considered as being based on one concept, that of a "state-description"... All other concepts can be defined in terms of this one; for example, the range of a formula is the class of state-descriptions whose disjunction is equivalent to the formula, a formula is analytic if its range contains all state-descriptions, and so on. Each state-description describes a "possible world" (described in as much detail as possible in our language). Thus we see that the range of a formula tells us in which possible worlds the formula would be true, and we see that an analytic formula is one which is true in all possible worlds (Kemeny 1952, 39).[26]

The concept of state-description opens up a way for a definition of an effective measure function upon a language system, which would assign a numerical value from the interval [0, 1] to every sentence of this language system. Still, there is a marked disadvantage of this procedure. We have already traced two extra-logical assumptions that Carnap accommodated in the construction of language system. The concept of state-description is accompanied by further extra-logical assumptions imposed upon language system. One of them concerns the individual constants:

PROPOSITION 1.3 *Extra-logical constraint 3*

each of the individual constants of language system signifies one and only one of the individuals of U.

The other assumption concerns the primitive predicates:

PROPOSITION 1.4 *Extra-logical constraint 4*

the primitive predicates must be interpreted in such a way that they "designate ... properties or relations which are logically independent of each other" (Carnap 1950, 73).[27]

The extra-logical constraints 3 and 4 are necessary in order for each state-description to be logically independent of other state-descriptions.[28] In view of the effectiveness of the measure function Carnap introduces further constraints:

PROPOSITION 1.5 *Extra-logical constraint 5*

the number of the individuals in U is at most denumerably infinite.

And the "requirement of completeness":

PROPOSITION 1.6 *Extra-logical constraint 6* (*Finite Variety Assumption*)

the set of primitive predicates in L must be sufficient for expressing every qualitative attribute of the individuals in the universe of L (Carnap 1950, 73-74).

The latter constraint, which will henceforth be referred to as the "finite variety assumption" is not of Carnap's own origin. He quotes Keynes:

> We seem to need some such assumption as that the amount of variety in the universe is limited in such a way that there is no one object so complex that its qualities fall into an infinite number of independent groups (Keynes 1921/1948, 258).

Commenting on the finite variety assumption Carnap concedes that it might not seem plausible with respect to qualitative predicates we ordinarily use, e.g. the qualitative visual predicates. Nonetheless,

> if... we think of the world as conceived in theoretical physics, then we find a finite and, indeed, a very small number of fundamental magnitudes to which, according to the assumption of physicists, the great variety of phenomena is ultimately reducible (Carnap 1950, 75).

The concept of the state-description finds one of its most important applications in yielding the normal form of the sentences of a language system. The other important application of this concept is in defining the concept of structure description (see definition 1.12 later in this section), which in Carnap's view becomes the proper domain of the prior probability distribution. The normal form then, via the concept of range of a given sentence, lets us determine the class of state-descriptions to which a given sentence is equivalent. The concept of the range of a given sentence is defined thus:

DEFINITION 1.3 *The range of a given sentence*

The range *of a given sentence p is the class of all the state-descriptions in which p holds* (Carnap 1950, 79).

If we follow the standard rules for defining first order logic, then it might be said that a given sentence p of L holds in a state-description S_1 of L iff:

1. p is an atomic sentence and it occurs as a conjunctive component of S_1; or
2. p is a *tautology of L; or*
3. *p is equivalent to* $\neg q$*, and q does not hold in* S_1*; or*
4. *p is equivalent to* $q \vee r$*, and at least one of the disjuncts holds in* S_1*; or*
5. *p is equivalent* to $q \wedge r$, and both conjuncts hold in S_1; or
6. p is equivalent to $\forall x P(x)$, and all instances of $P(x)$ in L hold in S_1.

Let me refer to the previous example of language system LS_3. Suppose we are interested in the range of the following sentence:

$$\neg P(a) \rightarrow P(b) \wedge P(c). \qquad (p_1)$$

First, note that p_1 is equivalent to the conjunctive form:

$$[P(a) \vee P(b)] \wedge [P(a) \vee P(c)].$$

Thus the range of p_1 includes 5 out of 8 state-descriptions available in a language system LS_3, i.e. p_1 does not hold in any of the following three state-descriptions:

$$\neg P(a) \wedge \neg P(b) \wedge P(c)$$

$$\neg P(a) \wedge P(b) \wedge \neg P(c)$$

$$\neg P(a) \wedge \neg P(b) \wedge \neg P(c);$$

and p_1 holds in all the others.

The concept of range is fundamental for the definition of the degree of confirmation, and for setting the fundamentals of Carnap's system of inductive logic. For consider the fact that once the language system is determined, so is the total number of state-descriptions of the language system, as well as the number of state-descriptions in which a given sentence of the language system holds for every sentence of the language system. The simplest solution now would be to assume that every possible world, i.e. every state-description, is equally possible, and thus for every state-description in language system it will hold that:

PROPOSITION 1.7 *Equally possible state-descriptions*

$$p(S_i) = 1/n;$$

where n is the total number of state-descriptions in a language system. This solution, however, is unsatisfactory for both Carnap and Leibniz (Hacking 1975, 52).

Let me recall now two important historical observations about *Logical Foundations of Probability*. First, Carnap—despite his declarations to the contrary—does not construe in the book an effective system of quantitative inductive logic. He relegates the task to the second volume, and only manages to sketch in the appendix how he intends his system to incorporate numbers, i.e. how precisely the logical relations between sentences of a given language system determine the measure function of these sentences. Second, in his sketchy remarks Carnap seems to be quite clear on the point that the logical measure function picks out precisely *one* confirmation function: "our final aim will be to choose *one* particular *c*-function as our explicatum" (italics mine; Carnap 1950, 483-84); a belief that he no longer sustained in *The Continuum of Inductive Methods* (Carnap 1952b), the intended follow-up to *Logical Foundations of Probability*.

Carnap projects the former book to serve as a preliminary conceptual setting of the logico-linguistic framework for his system of inductive logic as well as a detailed discussion of the criteria of adequacy for any quantitative theory of inductive logic. After having discussed several issues on inductive logic, e.g. the preceding qualitative and quantitative theories thereof, the problem of its application and relation to deductive logic, Carnap proceeds to a specification and detailed discussion of the criteria of adequacy (Carnap 1950, 286-95). The following are set out as elementary conventions:

PROPOSITION 1.8 *Equivalence*

a) *If e and e' are logically equivalent, then*

$$c(h \mid e) = c(h \mid e').$$

b) *If h and h' are logically equivalent, then*

$$c(h \mid e) = c(h' \mid e).$$

PROPOSITION 1.9 *Multiplication*

$$c(h \wedge j \mid e) = c(h \mid e) \cdot c(j \mid e \wedge h).$$

PROPOSITION 1.10 *Addition*

If $\neg(h \wedge j \wedge e)$ *is logically true, then*

$$c(h \vee j \mid e) = c(h \mid e) + c(j \mid e).$$

PROPOSITION 1.11 *Normalization*

If e is not logically false, then

$$c(t \mid e) = 1,$$

where t *is a tautology.*

PROPOSITION 1.12 *Initial confirmation*

If S_i is state-description that is not logically false and is finite, then

$$c(S_i \mid t) > 0.$$

The equivalence conditions (proposition 1.8) seem trivial, since they admit that the mere difference in linguistic form of the evidence or hypothesis statements does not affect the value of confirmation.[29] Multiplication (proposition 1.9) and addition (proposition 1.10) are—Carnap admits—"generally accepted in practically all modern theories of probability" (Carnap 1950, 285). The antecedent in the conditional of proposition 1.10 might be interpreted as implying that the two hypotheses are logically incompatible. Normalization imposes a limit on the maximal value of the confirmation function; together with the condition in addition it also means that the values of confirmation function range from 0 to 1 (endpoints included). Since $c(h \mid e) + c(\neg h \mid e) = 1$, it follows from proposition 1.10 (with $\neg h$ for j; and $t \equiv h \vee \neg h$) and proposition 1.11 that the minimal value of the confirmation function is $c(\neg t \mid e) = 0$. From the point of view of the concept of confirmation the values 0 and 1 are purely conventional;[30] nevertheless, they are non-arbitrary as far as other interpretations of the function $c(. \mid .)$ are involved, i.e. as the fair betting quotient and as the estimation of relative frequency (Carnap 1950, 286).

The conditions are almost equivalent to the standard probability axioms (Axiom 1)-(Axiom 3).[31] The only difference—apart from conditional probability—is in the addition principle, which in the case of the standard probability theory is stated in a general form:

PROPOSITION 1.13 *Addition—general*

$$c(h \vee j \mid e) = c(h \mid e) + c(j \mid e) - c(h \wedge j \mid e).$$

Remarkably, Carnap comments on the conditions specified as mere conventions, which are not sufficient to warrant the existence of "an adequate quantitative explicatum" of the logical concept of probability, but which have to be satisfied in the case of there being an adequate quantitative explicatum.

Let $m(.)$ be a "null confirmation" defined as follows:

DEFINITION 1.4 *Null confirmation*

$$m(p) = c(p \mid t)$$

From this definition, proposition 1.8 a) and proposition 1.9 it follows:

PROPOSITION 1.14 *Reduction of* $c(.\,|\,.)$ *to* $m(.)$

$$c(h \mid e) = c(e \wedge h \mid t) \,/\, c(e \mid t) = m(e \wedge h) \,/\, m(e);$$

Thus—via proposition 1.14—the task of finding the value of the confirmation function is reduced to finding the value of a measure function *m*(.). Since in any language system a given sentence is equivalent to a disjunction of state-descriptions, the problem of assigning a numerical value to each sentence of the language system reduces to the problem of assigning numerical values to each state-description of the language system, for the value of *m*(.) for a given sentence *p*—by addition (proposition 1.10)—is equivalent to the sum of the values of *m*(.) for each of the state-description in the range of *p*.

In *Logical Foundations of Probability* Carnap does not consider a uniquely determined confirmation function. He seems rather to attempt at narrowing down the class of confirmation functions which satisfy the adequacy conditions stated thus far. There are three important constraints discussed by Carnap in great detail: the regularity condition, positive relevance and the symmetry condition.

Carnap defines regularity as follows (Carnap 1950, 295):

DEFINITION 1.5 *M-Regularity*

m(.) *is* a regular measure function *for a language system iff m*(.) *fulfils the conditions:*

- *for every state-description* S_i *in a language system,* $m(S_i) > 0$*;*
- *the sum of the values of m*(.) *of all state-description of language system is 1.*

DEFINITION 1.6 *C-Regularity*

$c(p \mid q)$ (*where* $c(p \mid q) = m(p \wedge q) \,/\, m(q)$) *is* a regular confirmation function *iff m*(.) *is a regular measure function.*

Among the features of the regular confirmation functions, Carnap demonstrates the following counterparts of the axioms and theorems of the standard probability theory:

PROPOSITION 1.15 *Non-negativity for confirmation*

$$0 \leq c(h \mid e) \leq 1$$

PROPOSITION 1.16 *Maximum value for confirmation*

If it is logically true that e → h, then

$$c(h \mid e) = 1$$

PROPOSITION 1.17 *Normalization for confirmation*

$$c(t \mid e) = 1$$

PROPOSITION 1.18 *Addition for confirmation*

$$c(h \vee j \mid e) = c(h \mid e) + c(j \mid e) - c(h \wedge j \mid e)$$

PROPOSITION 1.19 *Bayes's theorem for confirmation*

$$c(h|e \wedge i) = \frac{c(h|e) \cdot c(i|e \wedge h)}{\sum_{j=1}^{n} c(h_j|e) \cdot c(i|e \wedge h_j)};$$

where $h_1, \ldots, h_j$ are mutually exclusive hypotheses and include h.

While Carnap appreciates that the theorem has been controversial, but he rebuts the objections as immaterial. First, the controversies are due to the confusion of logical and frequency interpretations of probability. Second, the controversies are due to the misapplication of the theorem on the basis of an uncritically understood principle of indifference; third, the theorem is proved on the assumptions specified thus far—the are weak assumptions "which practically all theories of probability seem to have in common" (Carnap 1950, 331). Finally, the theorem is also proved in the standard axiom system for probability.

In almost all theories of confirmation there is a desideratum that the accumulation of evidence compatible with the investigated hypothesis increases its degree of confirmation, i.e. this evidence is *positively relevant* to the hypothesis in question:

DEFINITION 1.7 *Positive relevance for confirmation*

j is positively relevant *to h on evidence e iff*

$$c(h \mid e \wedge j) > c(h \mid e).$$

This definition bears on the measure function $m(.)$ as follows:

PROPOSITION 1.20 *Positive relevance for measure function*

j is positively relevant *to h on evidence e iff*

$$m(h \wedge e \wedge j) \cdot m(e) > m(h \wedge e) \cdot m(e \wedge j).$$

Of the many candidates for confirmation function, the ones relying directly on the classical definition of probability – and thereby on the notorious principle of indifference – characteristically fail to satisfy positive relevance.[32]

Furthermore, the class of the regular confirmation functions is narrowed down considerably by symmetry conditions:

DEFINITION 1.8 *M-Symmetry*

m(.) is a symmetrical measure function *for a language system LS iff m(.) is a regular measure function for LS and for any expressions* φ, ψ *of language system if* φ *is isomorphic to* ψ, *then* $m(\varphi) = m(\psi)$.

Two expressions are *isomorphic* in a language system, roughly speaking, iff they differ only with respect to individual constants that occur within them. More precisely:

DEFINITION 1.9 *Correlation*

Crr is a correlation function *iff it is a one-one relation which maps the set of individual constants of the language system onto itself.*

DEFINITION 1.10 *Isomorphism*

Two expressions φ *and* ψ *of a language system are* isomorphic *iff there is such a Crr that* $Crr(\varphi)$ *and* ψ *differ at most in the order of conjuncts.*

Similarly for $c(. \mid .)$:

DEFINITION 1.11 *C-Symmetry*

$c(.\,|\,.)$ *is* a symmetrical confirmation function *for a language system iff* $m(.)$ *is a symmetrical measure function;*

where $c(p\,|\,q) = m(h \wedge e) \,/\, m(e)$.

Some of the symmetry properties of the $m(.)$ and $c(.\,|\,.)$ functions (with φ as Crr(ψ) and υ as Crr(δ)) are given below:

PROPOSITION 1.21 *Symmetry 1*

$$c(\psi\,|\,\delta) = c(\varphi\,|\,\upsilon).$$

PROPOSITION 1.22 *Symmetry 2*

if φ *and* υ *have no individual constants in common; likewise* ψ *and* υ*, then* $c(\varphi\,|\,\upsilon) = c(\psi\,|\,\upsilon)$.

The concept of state-description is a basic building-block of language systems. Carnap characteristically insists though that to convey the logical *structure* of language systems a different concept is needed:

DEFINITION 1.12 *Structure-description*

If Str *is a disjunction of all state-descriptions isomorphic to* S_i *in a language system LS, then* Str *is* a structure description *in LS.*

Given the above definition, the following symmetry theorems hold for structure-descriptions:

PROPOSITION 1.23 *Symmetry 3*

$$m(Str_j) = s \cdot m(S_i);$$

where s *is the number of state-descriptions isomorphic to* S_i *in the language system.*

PROPOSITION 1.24 *Symmetry 4*

$$c(S_i\,|\,Str_j) = 1/s.$$

The analysis of the regularity, relevance and symmetry conditions is taken by Carnap to constitute the substance of *Logical Foundations of Probability*. Nonetheless, these conditions define a much wider class of confirmation functions than Carnap would perhaps expect at the outset. However, in the *Appendix* (Carnap 1950, 562-77) he gives an outline of the projected confirmation function c^*:

DEFINITION 1.13 *Confirmation function c**

$$c^*(h \mid e) = m^*(h \wedge e) \,/\, m^*(e).$$

DEFINITION 1.14 *Measure function m**

m(.) is m(.) iff*

- *m(.) is a symmetrical;*
- *m(.) assigns the same numerical value to all structure-descriptions in the language system.*

If r is the number of structure-descriptions in the language system, it follows that:

PROPOSITION 1.25 *Symmetry 5*

$$m^*(S_i) = 1/r{\cdot}s.$$

There is still another option that Carnap brings to the fore:

DEFINITION 1.15 *Confirmation function c†*

$$c^\dagger(h \mid e) = m^\dagger(h \wedge e) \,/\, m^\dagger(e).$$

DEFINITION 1.16 *Measure function m†*

m(.) is m†(.) iff

- *m(.) is symmetrical;*
- *m(.) assigns the same numerical value to all state-descriptions in the language system.*

PROPOSITION 1.26 *Symmetry 6*

$$m\dagger(S_i) = 1/v;$$

where v is the number of all state-descriptions in the language system.

The two functions m^* and $m\dagger$ are both based on symmetry considerations. Only the former, however, could be satisfactory to Carnap, since the latter—based on the classical form of the principle of indifference—leads to contradictions, and also to the following result:

PROPOSITION 1.27 *Irrelevance*

$$c(h \mid e) = c(h \mid t).$$

This result seems, from the point of view of inductive inference, counter-intuitive, for all empirical evidence is irrelevant to both confirmation and prediction.

Several important changes in Carnap's views on confirmation and in his overall approach to the problem appeared first in the follow-up volume of the *Logical Foundations of Probability*, namely in *The Continuum of Inductive Methods*. He still sticks to the semantical framework of the construction, but the connection is loosened by the introduction of the concept of *the characteristic function*. It is a three-argument function $G(s, s_i, \kappa)$ and its values are in the interval [0, 1]. The first argument s is the number of the individuals—which are classified among the Q-predicates—observed in the sample. The second argument s_i is the number of individuals with the property Q_i among the observed individuals in the sample. The third argument κ specifies the number of Q-predicates in the language system ($\kappa = 2^\pi$; where π is the number of the primitive predicates in the language system).

The characteristic function is such as to give the *complete* and *unique* characteristics of a given confirmation function within the specified language system. The characteristics are complete if the value of the confirmation function is determined by the characteristic function for any pair of sentences within the language system. It is unique if for any pair of sentences of the language system the confirmation function has only one value. Carnap makes the value of the characteristic function dependent upon *the logical factor*, expressed here by κ, and *the empirical factor*, expressed by s and s_i (Carnap 1952b, 15-28). He supports this choice by the following argument. All thus far proposed confirmation functions lie between two extremes: on the one hand, the value of the confirmation function is totally independent of the empirical factor and is based exclusively on logical considerations; on the other hand, the confirmation function is nothing but a function of empirical data. The first proposal is Laplace's famous principle of indifference, which assigns equal values to equiprobable cases, i.e. cases with respect to which there is no reason to differentiate. In Carnap's terms the extreme confirmation function is a function of κ alone:

$$c(h_i \mid e) = 1/\kappa.$$

The probability of the hypothesis h_i that the next observed individual will instantiate Q_i depends only on the number of κ, and is the same for all κ Q-predicates, i.e. equals $1/\kappa$. The information carried in the evidence statement e is totally irrelevant to the value of the confirmation function.

On the other hand, the logical factor might be completely ignored in virtue of the empirical one:

$$c(h_i \mid e) = s_i / s.$$

The probability that the next observed individual will instantiate Q_i depends only on the ratio of the number s_i of thus far observed individuals instantiating Q_i to the total number s of the observed individuals in the sample.

Carnap claims that all thus far proposed confirmation functions may be located between the two extreme proposals. However, there apparently are good reasons to reject each of the two as inadequate for the characteristic function. The extreme logical confirmation function totally neglects the actual inferences carried out by scientists, who regard the empirical data as highly relevant to the confirmation of the conclusions. This method then is totally at odds with what is going on in science, and as such is in principle unacceptable as a confirmation function. Moreover, Carnap appeals to the well known fact that too-coarse formulations of this extreme form of the confirmation function would lead to contradictions (Carnap 1952b).

The extreme empirical proposal, on the other hand, is not free from paradoxical conclusions, either. Suppose, for instance, that one swan has been observed to be white. Then $s = 1$, $s_i = 1$, and

$$c(h_i \mid e) = s_i / s = 1.$$

On the basis of this extreme form of confirmation function it is not only true that the probability that the next observed swan will be white is 1, but also it is true that the probability that all swans are white is 1.

Moreover, it is unacceptable that the empirical extreme confirmation—if interpreted as a betting quotient—in similar cases would lead to bets with the apparently irrational odds 1:0, i.e. bets with a possible loss, but without a possible gain.[33]

Another reason for Carnap to reject this extreme empirical confirmation function is that it ignores the available logical information. Suppose, for instance, that two individuals have been observed, one of which instantiates P_1, the other P_2. On this proposal, however, the probability of the next observed individual instantiating property $P_1 \wedge P_2$ is 0. This is objectionable for Carnap, since such a statement is logically compatible with the thus far obtained information, and should not be excluded out of hand, however low its probability (Carnap 1952b, 42-43).

These considerations lead Carnap to set up the characteristic function in such a way that the two extreme proposals become its special cases, while it also contains a continuum of more moderate options. The characteristic function becomes thus a weighted mean of the logical ($1/\kappa$) and empirical (s_i/s) factors[34]:

DEFINITION 1.17 *The characteristic function G of the λ-system*

$$G(s, s_i, \kappa) = \zeta\cdot(1/\kappa) + (1-\zeta)\cdot(s_i/s) = (s_i + \lambda / \kappa) / (s + \lambda);$$

where $\zeta = \lambda / (s + \lambda)$.

The parameter λ may be treated either as a numerical constant or as a function of κ. In both cases its value is a non-negative real number, ranging from 0 to ∞. In the case $\lambda = 0$ we obtain the extreme empirical confirmation function; while in the case $\lambda \to \infty$ we obtain the extreme logical confirmation function. For all other non-negative values of λ we obtain confirmation functions which—to a different degree—take into account both factors, i.e. the logical and empirical.

There are several possible philosophical interpretations of the parameter. Hintikka, for instance, classifies them into subjective and objective. On the subjective interpretation λ is the index of caution: the higher its value the less a given person is ready to rely on a mere frequency count. On the objective interpretation λ depends on the order of the universe.[35] Hintikka's point then is that on none of these λ is a "logical" parameter, and extra-logical considerations have to be taken into account in order to account for the actual value of λ. This is also the case, he claims, with the α parameter he himself introduced into inductive logic (Hintikka 1997, 317).

Since the value of the parameter λ plays a crucial role in determining the ultimate shape of the confirmation function, Carnap decided to name his system of inductive logic referring to the name of the parameter: "the λ-system of inductive logic". In the ensuing I will follow this convention, and refer to the above system as "the λ-system". We obtain Carnap's earlier confirmation functions *c*† and c^* by suitable substitutions for λ, i.e. ∞ and κ respectively.

I proceed now to the examination of several remarkable divergences between the above discussed construction of inductive logic in *Logical Foundations of Probability* and *The Continuum of Inductive Methods*. I will focus on these divergences which are pertinent to the epistemological analysis pursued in the next chapters of this book.

Apparently, in the former book the construction of inductive logic is based on semantical concepts such as range or state-description. Moreover, it embraces Carnap's conviction that logical considerations alone will suffice to establish the set of adequacy conditions, which will uniquely determine the confirmation function. Both these essential convictions seem to have been given up in the latter book.

In order to specify the confirmation function Carnap in *The Continuum of Inductive Methods* introduces the concept of the characteristic function. The difference between the two concepts lies in the fact that the former is not a mathematical function, because it has propositional arguments, while the latter is a mathematical function. He apparently no longer sustains the view that logical considerations alone are sufficient to completely and uniquely determine the confirmation function. The semantical concepts, especially the concept of state-description, are no longer essential in determining the value of the confirmation function of the sentences of a given language system. Rather, the semantical concepts prove crucial in establishing the proof of the fact that the characteristic function gives a complete and unique characteristic of the confirmation function, i.e. gives one and only one value for any pair of sentences of the language system. Nevertheless, in other parts of the construal the semantical concepts seem superfluous.

The other important conviction that Carnap gave up in *The Continuum of Inductive Methods* concerns the possibility of determining a *unique* confirmation function. Definition 1.17 of the characteristic function clearly demonstrates that Carnap found it untenable that there is only *one* confirmation function, for there are as many functions satisfying this definition as many numerical constants can be substituted for the parameter λ occurring in definition 1.17. And since the value of the parameter λ ranges over the interval [0, ∞], this definition in fact determines the continuum of characteristic functions. Carnap is aware of the importance of this change for his overall approach to inductive logic:

> it would be conceivable that at some future time, on the basis of deeper insight, all will agree that a certain inductive method is the only valid one. On the other hand, it may be that the multiplicity of mutually incompatible methods is an essential characteristic of inductive logic, so that it would be meaningless to talk of "the one valid method". This does not necessarily imply that the choice of an inductive method is merely a matter of whim. ... However, questions of this kind would then not be purely theoretical but rather of a pragmatic nature. A method would here be judged as a more or less suitable instrument for a certain purpose (Carnap 1952b, 7).

Thus, instead of a single confirmation function for all inductive inferences, we have an infinite continuum of functions, with no specific clue as to the criteria of choice among them. Carnap recognized the difficulty, exposed later on by Salmon (Salmon 1967a), and tried to obstruct possible objections by claiming that the choice of the confirmation function is not a matter of an assertion, based on theoretical investigation, but a matter of a practical decision. One can thus change the confirmation functions as one can change "a saw or an automobile, and for similar reasons" (Carnap 1952b, 55).

Despite the change of his views with respect to there being only one confirmation function, Carnap insists on the fact that:

> For any given universe there is only one adequate language system or at least only one adequate value of κ (Carnap 1952b, 48).

The question arises: why is he still so confined to the idea of there being a uniquely determined language system for each domain of discourse? The motivation in *Logical Foundations of Probability* was rather perspicuous: each language system is supposed to uniquely determine its accompanying confirmation function, while what constitutes the major difficulty is to figure out the appropriate language system for each domain of discourse and investigate whether there is ultimately only one language system for each domain of discourse. Since in *The Continuum of Inductive Methods* the language systems's parameters do not uniquely determine the confirmation function what is the point in insisting on there being a one to one correspondence between different domains of discourse and the language systems? Why not refrain from determining a specific language system inasmuch as picking out a confirmation function is a matter of choice? These questions, I would suggest, find their appropriate answer only in the epistemological background of the systems of inductive logic, I will discuss in detail in the next chapter.

The later advancements in inductive logic were mostly stimulated by objections to Carnap's systems. Suppose we start with the following languages: L_1 with only one primitive predicate G for "green", and L_2 with two primitive predicates: R for "red", and Y for "yellow". Let the universe of discourse consist of two individuals signified by two individual constants a and b occurring both in L_1 and L_2. We need to begin with setting out the confirmation of a hypothesis prior to available evidence ($s = s_i = 0$):

PROPOSITION 1.28 *Prior confirmation*

$$c(h_i | e) = (s_i + \lambda / \kappa) / (s + \kappa) = 1/\kappa;$$

where h_i *is the predictive hypothesis that the next unobserved individual instantiates* Q_i.

In form of chart 1, below, I summarized all values necessary to carry out the calculations to follow (for simplicity assuming $\lambda = \kappa$).

Chart 1. State-descriptions and *structure-descriptions* for the languages $\boldsymbol{L_1}$ and $\boldsymbol{L_2}$.

L_2	*Structure description*		*State description*		*a*	*b*	*State description*		*Structure description*		L_1
	Weight	*No.*	*Weight*	*No.*			*No.*	*Weight*	*No.*	*Weight*	
Q_1: $R(x)Y(x)$; Q_2: $\neg R(x)Y(x)$; Q_3: $R(x)\neg Y(x)$; Q_4: $\neg R(x)\neg Y(x)$	1/10	I	1/10	1.	Q_1	Q_1	1.	1/3	I	1/3	Q_1: $R(x)$; Q_2: $\neg R(x)$
	1/10	II	1/10	2.	Q_2	Q_2	2.	1/3	II	1/3	
	1/10	III	1/20	3.	Q_1	Q_2	3.	1/6	III	1/3	
			1/20	4.	Q_2	Q_1	4.	1/6			
	1/10	IV	1/20	5.	Q_1	Q_3					
			1/20	6.	Q_3	Q_1					
	1/10	V	1/20	7.	Q_1	Q_4					
			1/20	8.	Q_4	Q_1					
	1/10	VI	1/20	9.	Q_2	Q_3					
			1/20	10.	Q_3	Q_2					
	1/10	VII	1/20	11.	Q_2	Q_4					
			1/20	12.	Q_4	Q_2					
	1/10	VIII	1/20	13.	Q_3	Q_4					
			1/20	14.	Q_4	Q_3					
	1/10	IX	1/10	15.	Q_3	Q_3					
	1/10	X	1/10	16.	Q_4	Q_4					

On the right-hand side of the chart is the specification of state-descriptions and structure-descriptions and their respective weights for the language L_1; on the left-hand side the specifications and weights for L_2 (notice the difference in the meaning of the Q-predicates in L_1 and L_2.) Since, as was demonstrated earlier, the specific values are sufficient for a definition of the confirmation function, we commence with the single-case predictive hypothesis to the effect that the next observed individual a will instantiate $R(x)$.

We begin with determination of the values of the confirmation function for the hypothesis $R(a)$ prior to any observation of a sample. In L_1 $R(a)$ is equivalent to the disjunction of the state-descriptions 1 and 3 of L_1. Thus:

$$c_1(R(a) \mid t) = c_1(Q_1(a)Q_1(b) \vee Q_1(a)Q_2(b) \mid t).$$

Since the state-descriptions 1 and 3 of L_1 are logically exclusive, from the special addition condition (proposition 1.10) it follows that:

$$c_1(Q_1(a)Q_1(b) \vee Q_1(a)Q_2(b) \mid t) = c_1(Q_1(a)Q_1(b) \mid t) + c_1(Q_1(a)Q_1(b) \mid t).$$

By inserting the appropriate values from chart 1 we obtain the following value for the prior confirmation in L_1:

$$c_1(R(a) \mid t) = 1/3 + 1/6 = 1/2. \qquad \text{(PC1)}$$

By similar calculation for L_2, where the hypothesis $R(a)$ is equivalent to the disjunction of the state-descriptions 1, 3, 5, 6, 7, 10, 13, and 15 of L_2, we obtain:

$$c_2(R(a) \mid t) = 2/10 + 6/20 = 1/2. \qquad \text{(PC2)}$$

This intuitively plausible equivalence of the values of the prior confirmation in L_1 (PC1) and in L_2 (PC2) will change, however, as soon as we consider the observed sample. Suppose that, independently of the language we are using, the evidence statement e reports that one object has been observed to be R. In order to compute the values of the confirmation function we need to take notice that the values of κ differ in the case of L_1 ($\kappa = 2$) and L_2 ($\kappa = 4$). By application of definition 1.17 we obtain the following values of confirmation of $R(a)$ conditional on evidence e:

$$c_1(R(a) \mid e) = (1+1) / (1+2) = 2/3. \qquad \text{(PE1)}$$

$$c_2(R(a) \mid e) = (1+1) / (1+4) = 2/5. \qquad \text{(PE2)}$$

The same sentence, occurring though in two different languages L_1 and L_2, obtains then not only two different values of confirmation by the same evidence: (PE2) < (PE1), but also the increase of its degree of posterior confirmation with respect to prior confirmation is accordingly different:

$$(\text{PE2}) - (\text{PC2}) < (\text{PE1}) - (\text{PC1}).$$

The fact that the confirmation value of a given hypothesis depends on the number of predicates used in a given language seems highly unintuitive from the point of view of scientific practice. In other words, Carnap's systems apparently fail to satisfy the criterion of linguistic invariance:

> Let $c(h \mid e)$ be a confirmation function defined for languages $L_1,\ldots, L_r$. Let h_i and e_i be two sentences of L_i ($1 \leq i \leq r$) such that $c(h_i \mid e_i) = c_1$ in L_i. To satisfy the criterion of linguistic invariance the confirmation function $c(h \mid e)$ must be such that it is impossible, using the values of $c(h \mid e)$ for any sentences h and e in any of the languages $L_1,\ldots, L_r$ and using the semantical and syntactical rules of $L_1,\ldots, L_r$, to prove that $c(h_i \mid e_i) = c_2$, where $c_1 \neq c_2$ (Salmon 1961, 256).[36]

In order to arrive at the values (PC1) and (PC2) we applied the special addition condition to single out state-descriptions. However, the latter are logically exclusive only if the primitive predicates they consist of are logically independent, i.e. the ascription of one of them to a given object does not determine the truth-value of the ascription of another one to the same object. Otherwise, the state-descriptions would include self-contradictory sentences, each of which being assigned a non-negative prior confirmation value (the regularity conditions), and 0 (the probability axioms). The assumption of logical independence of primitive predicates is an extra-logical assumption, which is not only very restrictive with respect to a chosen language, but also appears to be arbitrary (Nagel 1963).

Not surprisingly then, the first accomplishments of Carnap's systems of inductive logic were oriented towards relaxation of these extra-logical and restrictive assumptions. The most successful attempts are due to J. G. Kemeny:

> many restrictions hitherto imposed have been removed. It is not necessary that L be a first order calculus, it is not necessary that all individuals have names, it is not necessary that the constants be logically independent of each other, and it is not necessary that every qualitative attribute be expressible in L. These are four steps in the direction of widening the applicability of our general conditions of adequacy... (Kemeny 1963, 718).

In sum, Kemeny's improvements over Carnap's initial versions of the systems of inductive logic not only relax the language bonds of the former;

they proved extremely important in allowing the later developments of the systems into an axiomatic form and also their full mathematization.[37]

Kemeny's idea is roughly as follows. Instead of Carnap's favorite semantical concept of state-description, he bases the definition of the semantical measure-function on the concept of model. Every axiom of the system has a set of models, in which it is true. The intersection of these sets is a set of models in which all axioms are true. The measure for the set of all models is 1, and 0 for an empty set. The measure is additive and non-negative; it also fulfills the definition of σ-measure (i.e. if defined for 2 sets, then also for their complements, sum and product).[38]

There is still another deficiency in Carnap's systems, which became a starting point for Hintikka and his co-workers in developing their own version of the systems. Let me illustrate it with the formerly discussed example of L_1 and L_2. Suppose we are interested in the confirmation value of a general sentence

$$\forall x R(x) \qquad \text{(GS)}$$

Due to the construction of the systems of inductive logic, in both languages (GS) is equivalent to a disjunction of singular sentences, namely those state-descriptions with which it is consistent. Thus, in L_1 (GS) is equivalent to the state-description 1:

$$\forall x R(x) \equiv Q_1(a)Q_1(b)$$

and the confirmation value of (GS) in L_1 prior to any experiment is:

$$c_1(\forall x R(x) \mid t) = c\big(Q_1(a)Q_1(b) \mid t\big) = 1/3. \qquad \text{(PCGS1)}$$

Similarly, in L_2 (GS) is equivalent to the state-descriptions 1, 5, 6 and 15. Hence,

$$c_2(\forall x R(x) \mid t) = 2/10 + 2/20 = 3/10. \qquad \text{(PCGS2)}$$

And, in general (assuming that there are N individuals in the domain of each language):

$$c_1(\forall x R(x) \mid t) = (1/N) \,/\, (1 + 1/N), \qquad \text{(GPCGS1)}$$

$$c_2(\forall x R(x) \mid t) = (6/N^2) \,/\, (1 + 5/N + 6/N^2). \qquad \text{(GPCGS2)}$$

Thus, if the number N of individual objects is infinite, we obtain that:

$$\lim_{N\to\infty} c_\kappa(\forall x R(x)|t) = \frac{\dfrac{u_{\kappa/2}}{N^{\kappa/2}}}{1+\dfrac{u_1}{N}+\dfrac{u_2}{N^2}+\ldots+\dfrac{u_{\kappa/2}}{N^{\kappa/2}}} = 0;$$

where $u_1 \ldots u_{\kappa/2}$ are polynomial parameters.

It is not only the case of such simple general sentences as (GS), but a central feature of Carnap's system of inductive logic that general sentences have zero confirmation in language systems with infinite N.[39]

3. The Finnish School's Systems of Inductive Logic

I follow here the standard exposition of the motivation for Hintikka's α-λ-system (cf. Mortimer 1982; Festa forthcoming), according to which the zero-confirmation problem was the starting point for the elaboration of a new version of inductive systems by Hintikka, accompanied later on by his co-workers.[40] However, it is debatable whether the zero confirmation for generalizations is indeed devastating. In fact, Carnap was the first to notice this feature of his λ-system, while he apparently failed to appreciate it as a major problem. Perhaps because it accords well with his views on the role of generalizations in science (Carnap 1950).[41]

Hintikka acknowledges that he continues the program initiated by Carnap

> of constructing a quantitative inductive logic (for the same kinds of language systems as Carnap) which avoids some of the main disadvantages of Carnap's theory in so far as the problem of dealing with strict (non-statistical) inductive generalization is concerned.
>
> Such an inductive logic can even be built along lines closely reminiscent of Carnap's. In developing an inductive logic along Carnapian lines the fundamental problem concerns the choice of the underlying measure function. ... Now one way in which one comes to choose the particular measure function m^* which Carnap favors may perhaps be described as follows: What we are primarily interested in are not the different "possible worlds" described by state-descriptions but rather different *kinds of* possible worlds that can be described in our language. These different kinds of worlds are interpreted by Carnap as structurally different worlds (Hintikka 1965b, 278-79).

I quote Hintikka's own acknowledgment,[42] for in a recent comment he repudiates the phrase "Carnap-Hintikka style systems of inductive logic"

(Hintikka 1997, 317) as inadequate for the epistemological divergence between him and Carnap. Despite this seeming functional discrepancy on epistemic grounds, there is, admittedly, the homogeneity of structure and origin.

The quotation also gives a rough approximation to the procedure employed by Hintikka. Carnap commenced with bringing out the conditions of adequacy and then tried to semantically (in terms of state-descriptions) ground the confirmation function in a way consonant with the conditions. Hintikka seems to sustain all unproblematic parts of Carnap's attitude, and overtakes the semantical framework. However, what he finds problematic about state-descriptions is that they are conjunctions of singular sentences which, in effect, results in the zero-confirmation problem. This is why in the above quotation he introduces the idea of a description of "kinds of worlds" instead of "possible worlds". And that in turn requires a somewhat different formal apparatus. Technically, the idea is elaborated on the basis of Hintikka's favorite normal form, i.e. the distributive normal form for the first-order predicate logic.[43] Roughly, the idea is that instead of predicating each of the individuals in U one of the Q_i-predicates, we list non-empty Q_i-predicates (called henceforth *attributive constituents*[44]) and add that these are all non-empty attributive constituents. The formula listing all attributive constituents instantiated in a given world is called *constituent*:

DEFINITION 1.18 *Constituent*

$$\exists x Ct_1(x) \wedge \exists x Ct_2(x) \wedge \dots \wedge \exists x Ct_i(x) \wedge \forall x\big(Ct_1(x) \vee Ct_2(x) \vee \dots \vee Ct_i(x)\big);$$

where Ct_i—is one of the $K = 2^{\pi}$ Q_i-predicates in a given language.

There are 2^K constituents in a given language:

> Constituents describe all the different kinds of "possible worlds" that can be specified by means of our monadic predicates, quantifiers and propositional connectives (Hilpinen 1966, 133-34).[45]

Carnap admits that constituents are more abstract semantical concepts than state-descriptions. However, he goes on to suggest that:

> we might carry this approach one step further. We can go from the constituents to what might be called constituent-structures. Instead of saying of each predicate whether it is exemplified or not, we simply say how many of them are instantiated in the world and how many are empty. ... We might then see what happens if we give an equal a priori probability to each of these constituent-structures (Carnap 1968c, 219).

The definition of confirmation function for constituents must, according to Hintikka, satisfy not only Carnap's but also two additional criteria of adequacy:

(1) Among constituents representing strong generalizations compatible with the evidence there will be only one such that its probability approaches 1 in the limit as the positive evidence increases.

(2) The simplest constituent compatible with evidence will have the greatest a posteriori probability (Tuomela 1966, 163).

It seems that these criteria lead naturally then to make Carnap's criteria for singular sentences conditional on the two criteria for general sentences. And this feature is in fact reflected in the very construction of the confirmation function in the system of inductive logic propounded by Hintikka.

As has been proved by Carnap in *The Continuum of Inductive Methods* the values of confirmation function for any pair of sentences of a given language are known as soon as the special value is known. Hintikka arrives at the special value (the probability that the next observed individual instantiates the attributive constituent Ct_i conditional on the evidence that s observed individuals instantiate c attributive constituents) via the prior probability of the evidence, which has the following form:

PROPOSITION 1.29 *Prior probability of evidence*

$$p(e_c)=\sum_{i=0}^{K-c}\left[\binom{K-c}{i}p(C_{c+i})\,p(e_c|C_{c+i})\right];$$

where e_c —is the evidence to the effect that s thus far observed individuals instantiate c attributive constituents.

The special value is determined as soon as the value of the prior probability $p(C_w)$ of C_w, i.e. a constituent which claims that of attributive constituents w are instantiated, as well as the likelihood $p(e_c \mid C_w)$ are determined.[46] On the one hand, Hintikka wants to make the value of $p(C_w)$ depend upon w only; on the other hand, he wants to hold fast to Carnap's way of determining the prior probabilities (Hintikka 1966, 115). According to the definition of the characteristic function G of Carnap's λ-system the probability that α individuals are compatible with a particular set of w attributive constituents is given by the formula:

PROPOSITION 1.30 *Probability of w instantiated constituents (λ-system)*

$$\frac{(w\lambda/K)}{\lambda}\frac{1+(w\lambda/K)}{1+\lambda}\cdots\frac{\alpha-1+(w\lambda/K)}{\alpha-1+\lambda}=\frac{\Gamma(\alpha+w\lambda/K)\,\Gamma(\lambda)}{\Gamma(\alpha+\lambda)\,\Gamma(w\lambda/K)};$$

where Γ(.) is the gamma function.

Since, however, Hintikka insists on infinite domains of individuals and also on the dependence on w only (which is analogous to the structural classification among state-descriptions in the λ-system) the value of $p(C_w)$ is proportional to probability of w instantiated constituents (proposition 1.30) with due adjustments:

PROPOSITION 1.31 *Probability of w instantiated constituents (α-λ-system)*

$$p(C_w)=\frac{\dfrac{\Gamma(\alpha+w\lambda(K)/K)}{\Gamma(w\lambda(K)/K)}}{\displaystyle\sum_{i=0}^{K}\binom{K}{i}\frac{\Gamma(\alpha+i\lambda(K)/K)}{\Gamma(i\lambda(K)/K)}};$$

where λ (*K*)—is a function of *K*; admittedly, in a simplified form λ may be just a numerical constant.

The special value of the λ-system is not directly applicable to the α-λ-system—mainly due to a different semantical framework, as has been already pointed out—but still, as Hintikka claims, there is room for suitable application thereof. There is a special value conditional on the fact that the constituent C_w is true, i.e. conditional on the fact that w attributive constituents are instantiated. This, then, naturally suggests, according to Hintikka, the following modification of the special value of the λ-system:

PROPOSITION 1.32 *The special value of the α-λ-system*

$$(s_j+\lambda/w)\,/\,(s+\lambda).$$

The substitution of w for κ is due to the fact that—conditional on C_w—it is presupposed that the number of attributive constituents is w, not κ. In this way we obtain the required likelihood:

PROPOSITION 1.33 *The likelihood for the α-λ-system*

$$p(e_c|C_w) =$$

$$\frac{(\lambda/w)}{\lambda}\frac{(1+\lambda/w)}{\lambda+w}\cdots\frac{s_1-1+\lambda/w}{s_1-1+\lambda}\cdots\frac{s_c-1+\lambda/w}{s-1+\lambda}=$$

$$=\frac{\Gamma(\lambda)}{\Gamma(s+\lambda)}\prod_{i=1}^{c}\frac{\Gamma(s_j+\lambda/w)}{\Gamma(\lambda/w)};$$

where s_1 is the number of observed objects instantiating Ct_1 ($s_1 + \ldots s_K = s$).

By relevant substitutions and by the definition of conditional probability:

$$p(h \mid e) = p(h \wedge e) / p(e)$$

which in the case of $p(Ct_j \wedge e_c)$ requires that the former s_j be changed to $s_j + 1$, and s to $s + 1$ (the formula states then that one additional individual is assumed to have been observed to be Ct_j) we obtain the following unconditional special value for Hintikka's α-λ-system:

PROPOSITION 1.34 *Unconditional special value for the α-λ-system*

$$p(Ct_j|e_c) =$$

$$\frac{\sum_{i=0}^{K-c}\binom{i}{K-c}\left(\frac{\Gamma(\alpha+(c+i)\,\lambda(K)/K)\,\Gamma(\lambda(c+i))}{\Gamma((c+i)\lambda(K)/K)\,\Gamma(s+1+\lambda(c+i))}\left[\prod_{j}^{c}\frac{\Gamma\left(s_j+\frac{\lambda(c+i)}{c+i}\right)}{\Gamma\left(\frac{\lambda(c+i)}{c+i}\right)}\right] \times \frac{\Gamma\left(s_j+1+\frac{\lambda(c+i)}{c+i}\right)}{\Gamma\left(\frac{\lambda(c+i)}{c+i}\right)}\right)}{\sum_{i=0}^{K-c}\binom{i}{K-c}\left[\frac{\Gamma(\alpha+(c+i)\,\lambda(K)/K)\,\Gamma(\lambda(c+i))}{\Gamma((c+i)\lambda(K)/K)\,\Gamma(s+\lambda(c+i))}\prod_{j=1}^{c}\frac{\Gamma\left(s_j+\frac{\lambda(c+i)}{c+i}\right)}{\Gamma\left(\frac{\lambda(c+i)}{c+i}\right)}\right]}$$

By appropriate application of Bayes's theorem we obtain the posterior probability of C_w:

PROPOSITION 1.35 *Posterior confirmation in the α-λ-system*

$$p(C_w|e_c) =$$

$$\frac{\frac{\Gamma(\alpha+w\lambda(K)/K)\,\Gamma(\lambda(w))}{\Gamma(w\lambda(K)/K)\,\Gamma(s+\lambda(w))}\prod_{j=1}^{c}\frac{\Gamma(s_j+\lambda(w)/w)}{\Gamma(\lambda(w)/w)}}{\sum_{i=0}^{K-c}\binom{K-c}{i}\left[\frac{\Gamma(\alpha+(c+i)\lambda(K)/K)\,\Gamma(\lambda(c+i))}{\Gamma((c+i)\lambda(K)/K)\,\Gamma(s+\lambda(c+i))}\prod_{j=1}^{c}\frac{\Gamma(s_j+\lambda(c+i)/(c+i))}{\Gamma(\lambda(c+i)/(c+i))}\right]}$$

The formulae in proposition 1.34, proposition 1.31 and proposition 1.35 give the basic characteristics of the α-λ-system. The meaning of the λ parameter with respect to singular sentences is the same as in the λ-system. There is, however, a new parameter: α. It is instructive to observe its influence upon the values of confirmation. If $\alpha \rightarrow \infty$, then the special value of the α-λ-system (proposition 1.34) is the same as the special value of the λ-system based on definition 1.17. Since the special value determines completely a given system of inductive logic, it follows that Carnap's λ-system is a special case of the α-λ-system.

On the other hand, if $\alpha = 0$, then the prior probabilities of C_w turn out to depend on K only in that "a highly *a priori* method of determining the probabilities $p(C_w)$" is set out (Hintikka 1966, 123). Nevertheless, $p(C_w \mid e_c)$ is in this case not so much affected by $p(C_w)$ as by the results of experiments. Hence, the general conclusion for the interpretation of α is:

> The larger this parameter is, the more suspicious we are of prospective inductive generalizations (Hintikka 1966, 126).[47]

For, as Hintikka claims, α is, similarly to κ, an index of the weight of the logical factor in inductive inferences: with increasing α the value of $p(C_w)$ for $w < K$ approaches 0. The main difference then lies in the fact that α controls the behavior of generalizations, while λ—of singular sentences (Hintikka 1966, 118).

Hintikka's approach initiated a number of contributions by his co-workers in accomplishing various aspects of the α-λ-system. Some of them concern technical advances, while others deal with epistemological issues.

Among the most significant technical improvements one should list R. Hilpinen's attempt to extend the language of the system to represent not only monadic, but also dyadic predicates (Hilpinen 1966; 1971). He elaborates the definition of the attributive constituent and the constituent suitable for this kind of language, and then proceeds to the definition of the confirmation function according to Hintikka's criteria of adequacy (1) and (2) listed above. In contrast, however, to language containing only monadic predicates it seems that there is, according to Hilpinen, no unique definition of the simplest generalization fitting the evidential input.[48] The important epistemological moral is that there is no clear-and-cut way to separate the factual and inferential aspects of knowledge as Bayesians would like to have it (Hilpinen 1971, 282; Audi 1998, 154-56). Carnap, for instance, believed that there is a sharp boundary between the semantical and analytic concept of the degree of confirmation, explicated as "$c(.\mid.)$", which constitutes the "inferential" aspect of knowledge, and its "factual" complement in the form of the synthetic sentences, which are the arguments of "$c(.\mid.)$", especially sentences representing evidence.

Tuomela was the first to attempt to represent order relations in the system (Tuomela 1966). The basic idea is to extend the language in such a way as to enable formulation of generalizations describing order in a given universe of discourse. He achieves this by introducing a dyadic predicate "immediate successor", and then by adjustments to the definition of attributive constituent and constituent suitable for this kind of language. His solution, however, is also relative to the definition of the simplicity of constituents for the defined kind of language.

In Hintikka's system old epistemological problems related to the inductive justification of generalizations reappear. For on the one hand, it is possible in the system that $p(C_w | e) > 0$, while the number of objects observed thus far s is less than w. On the other hand, according to Hintikka's own criteria of adequacy there are generalizations preferable from the point of view of evidence, i.e. such that their degree of confirmation increases faster than others. Hintikka tries to account for these features of the system in terms of "nomic" and "accidental" generalizations (Hintikka 1966, 121-22). The problem is subsequently dealt with in its full extent by J. Pietarinen (Pietarinen 1972). His argument is to the effect that every system of inductive logic hangs on the presuppositions concerning what the world is like, in particular what is the order in the universe of discourse. Having settled this, it is possible to provide the parameters α and λ with an appropriate interpretation so as to make them function as the measures of lawlikeness.

4. The Bayesian Developments of Inductive Logic

Broadly speaking, Bayesianism is a general and uniform account of scientific knowledge. It identifies the epistemic support evidence e confers on a hypothesis h with probability which is defined by axioms (Axiom 1)-(Axiom 3) (see page 7). More precisely, this epistemic support is identified with hypothesis's conditional probability on e, i.e. with $p(h | e)$. The latter is then calculated according to Bayes's theorem. The rule of *conditionalization* constitutes an integral part of Bayesianism. This rule prescribes how to update one's beliefs on changing evidence. If we acquire new evidence e_2, then the probability of h on the old evidence e_1 becomes a prior probability of h:

DEFINITION 1.1 *Rule of conditionalization* (C)

$$p(h | e_1) = q(h);$$

where in accordance with Bayes's theorem (see page 5):

$$q(h | e_2) = q(h) \cdot q(e_2 | h) / q(e_2).$$

An important step towards the development of the Bayesian reading[49] of the systems of inductive logic was due to Carnap's change of the logical framework for the systems; a change definitely inspired by Kemeny. In "A Basic System of Inductive Logic, Part I" (Carnap 1971a, 53-67).[50] Carnap defines the set of possible worlds in terms of models and propositions, and not in terms of syntactical combinations of *Q*-predicates. Nevertheless, as proved by T. Fine (Fine 1973, 193), the value of the confirmation function defined in this new setting still depends on the number of primitive predicates of a given language system.

The definition of the set of possible worlds enables Carnap to define a counterpart of the standard domain of probability measure, i.e. the σ-field. And this, in turn, proved influential with respect to providing the first axiomatic presentation of his λ-system.[51] Moreover, W. Stegmüller generalized the characteristic function of the λ-system and obtained the characteristic function of a new two-parametric γ-λ-system:

PROPOSITION 1.36 *The characteristic function GC of the γ-λ-system*

$$G(s, s_i, \kappa) = (s_i + \lambda \cdot \gamma_i) / (s + \lambda);$$

where γ_i is prior probability of Q_i's. In particular, if for all *Q*-predicates γ_i is equal, then $\gamma_i = 1/\kappa$ and GC is equivalent to G (see definition 1.17).

The α-λ-system was axiomatized by J. Hintikka and I. Niiniluoto in 1976.[52] It was, however, Kuipers, who developed a systematic mathematical framework for various systems of inductive logic by adopting the standards of probability theory and statistics. Notwithstanding the difference in means of representation,[53] the axiomatic versions of the systems aim at a generalization of the pre-axiomatic ones. The systems of inductive logic are obtained by appending to the standard probability axioms additional constraints, subsequent versions of which tend to be more general, allowing thus to derive preceding systems as a special case. Hence, on this account, every system of inductive logic satisfies the following standard axioms of probability calculus (Axiom 1)-(Axiom 3). Apart from these axioms the γ-λ-system satisfies the following principles:[54]

PROPOSITION 1.37 *Principle of order indifference*

$$p(Q_iQ_j \mid e_s) = p(Q_jQ_i \mid e_s).$$

PROPOSITION 1.38 *Principle of restricted relevance*

$$p(Q_i \mid e_s) = f(s, s_i).$$

PROPOSITION 1.39 *Principle of initial possibility*

$$\gamma_i > 0.$$

The principle of order indifference is an application of the principle of indifference with respect to Q-predicates. It was introduced by Carnap, but one needs to be mindful that in Kuipers's formulation we are no longer dealing with predicates, but rather *Q-sets*. The principle of restricted relevance makes the special value depend only upon s, i.e. the number of individuals observed thus far, and the number s_i of those among s, which fall within the investigated category Q_i.

All thus far discussed systems of inductive logic could be represented in a similar way, and also some other systems, which then prove to be special cases or generalizations of the foregoing.[55]

The conceptual modification[56] proved fruitful in stimulating the further development of systems of inductive logic, especially the Bayesian reading thereof. S. L. Zabell (Zabell 1982), for instance, demonstrated that the γ-λ-system, reformulated accordingly, is equivalent to the sufficientness postulate advanced by W. E. Johnson in Bayesian statistics.

The Bayesian reading of the γ-λ-system is as follows: suppose there are t possible categories $\{c_1, c_2, ..., c_t\}$, where $2 \leq t \leq \infty$. Let $X_1, X_2, X_3, ...$ be a sequence of exchangeable (i.e. invariant under finite permutations of outcomes)[57] random variables with values in the set of categories. Let $n_i = n_i(X_1, X_2, ..., X_n)$ for $n \leq 1$ be the number of outcomes within the sample $X_1, X_2, ..., X_n$ falling within a given category $X_k = c_i$, and let $\mathbf{n} = (n_1, n_2, ..., n_t)$ specify the list of the respective numbers for all the categories. The predictive probability (or Carnap's "special value"), i.e. the probability that the next observed individual will fall within the category c_i, conditional on the observed sample frequencies is:

PROPOSITION 1.40 *Predictive probability*

$$P(X_{n+1} = c_i \mid \mathbf{n}) = (n_i + \alpha_i) / (n + \alpha);$$

where $\alpha_i > 0$; $\alpha = \Sigma\alpha_i$; $n \geq 0$ (Zabell 1997, 267).

Apparently, if we let $\alpha_i = \lambda\gamma_i$, then the predictive probability is precisely the same as in the γ-λ-system (see also Skyrms 1993c, 79).

Johnson's sufficientness postulate leads to the above formula for predictive probability.[58] It makes the latter depend only upon the observed sample frequency for the category at hand:

PROPOSITION 1.41 *Sample frequency*

$$P(X_{n+1} = c_i \mid \mathbf{n}) = f_i(n_i).$$

In other words, the sufficientness postulate may be read as composed of two conditions:

(i) The probability of an outcome on the next trial should only depend on the frequency that *it* has occurred in the preceding trials (for each fixed number of trials) and not on the relative frequencies of the other outcomes.
(ii) The dependence should be the same for all categories (Skyrms 1993c, 81).

To obtain Carnap's λ-system on the Bayesian grounds we need the following: (1) exchangeability and not independence; (2) sufficientness; (3) $t \geq 3$; (4) strict coherence. If the condition (ii) of Johnson's postulate is dropped, then we obtain the γ-λ-system.

There are two alternative readings of the γ-λ-system on the grounds of Bayesian statistics: the objective reading and the subjective one. On the former the γ-λ-system corresponds to inferences in multinomial contexts, which are described as follows:[59] suppose we have $k \geq 2$ mutually exclusive and jointly exhaustive categories $c_1, \ldots, c_k$. They classify the outcomes of an experimental process E. It is further supposed that for each category there is a *physical probability* $\mathbf{q}_\mathbf{i}$ that the outcome of a trial in E falls within c_i, and this probability is supposed to be constant across E. We obtain thus a multivariate (or in the case $k = 2$ binomial) Bernoulli process. The standard examples of this kind of process are, for instance, the following: throwing a die with unknown—but unchanged across the trials—bias, or sampling with replacement from an urn of unknown composition (Skyrms 1996a, 323).

If the scientist investigating E assumes the latter to be a multivariate, in particular binomial, Bernoulli process, then we describe the scientist's cognitive context as a *multinomial context*. Since the scientist will probably not know the true values of $\mathbf{q}_\mathbf{i}$ for each category c_i ($1 \leq i \leq k$) she will try to investigate various hypotheses concerning the true values of $\mathbf{q}_\mathbf{i}$ on the grounds of the outcomes of E. The inferences, which lead to establishing how far the hypotheses are confirmed by the evidence, in this context will be called *multinomial inferences*.

In sum, we assume that there is an objective (physical) chance mechanism, and the prior distribution reflects the scientist's subjective uncertainty[60] with respect to which the probability vector $\mathbf{q} = (\mathbf{q}_1, \ldots, \mathbf{q}_k)$ governs the chance mechanism. The aim of inductive inference is to learn the true values of the vector $\mathbf{q}$.

On this objective reading it is worth noting that the prior distribution imposed within the γ-λ-system is equivalent to the prior *Dirichlet distribution* in Bayesian statistics (Skyrms 1993c, 81). For in the situation prior to any experiment (where $n_i = n = 0$) the predictive probability defined for the γ-λ-system gives us the complete prior distribution:

$$P(X_{n+1} = c_i \mid \mathbf{n}) = \alpha_i / \alpha = \gamma_i.$$

Thus determined prior distribution is equivalent to the Dirichlet distribution with the parameters λ and $\gamma = (\gamma_1, \ldots, \gamma_k)$, i.e. Dir(γ, λ) (Festa 1995, 303).

Not every Bayesian, however, will assent to the assumption that there is anything like *physical probabilities* described above. For subjective Bayesians contest any reference to the objective or physical chances of an outcome falling within a category (Skyrms 1993c, 81; 1996, 324). Rather, they consider the concept of probability as more or less adequately expressing our degrees of belief that an outcome will fall within a given category. What is sufficient to determine the predictor's degree of belief that the outcome of the next trial will fall within a category c_i is the observed frequency for this category. Johnson's postulate is then indispensable for this purely subjective Bayesian reading of the γ-λ-system. From this perspective, the conditions required to infer the γ-λ-system are interpreted as imposing strong symmetries among the predictor's beliefs rather than defining inductive inferences in multinomial contexts (Skyrms 1993c, 82).

The interpretation of Carnap's γ-λ-system on the grounds of Bayesian statistics proved extraordinarily fruitful for the former. A whole range of problems that Carnap recognized as significant and which he was not able to overcome, seems to have found natural solutions in the Bayesian perspective. Of these I will shortly discuss two problems of analogy,[61] the problem of the choice of the optimum inductive method,[62] and the problem of zero-confirmation of generalizations (see section 2 in this chapter).

A system of inductive logic of the Carnap-Hintikka style is completely characterized by its special value. Let us take the special value of the γ-λ-system in the form of predictive probability (proposition 1.40). Skyrms proposes a natural generalization of the system in question, which will be capable of representing the continuum of values (Skyrms 1993; 1996, 325-330). For in the form proposed by Carnap the γ-λ-system is capable of dealing only with non-negative integers. In the place of α in proposition 1.40 Skyrms introduces a parameter β, which is defined as a non-negative bounded measure of the Borel algebra of the unit interval.[63] Thus, if O is a Borel set in the unit interval and **n** is the observed frequency vector reporting n_i points in O in n trials, the predictive probability that the outcome of the next trial falls in O is:

PROPOSITION 1.42 *Predictive probability for the value continuum*

$$P(O \mid \mathbf{n}) = (\beta(O) + n_i) / (\beta([0, 1)) + n).$$

Skyrms gives the following illustration of how this formula generalizes predictive probability (proposition 1.40) to represent the "value continuum" (Skyrms 1993c, 83; 1996, 326). Suppose I is a subinterval of $[0, 1)$, and $\beta(I)$ is just the length of I. Suppose further that the unit interval is divided into k subintervals, that correspond to k categories of outcomes. The fact that of n trials n_i resulted in I, would be reported in terms of proposition 1.42 as follows:

$$P(I \mid \mathbf{n}) = (1/k + n_i) / (1 + n).$$

We can easily recognize that the above formula is a special value of Carnap's γ-λ-system (where it is assumed that $\lambda = 1$, and γ_i is uniform for all the categories).

Predictive probability for the value continuum (proposition 1.42) is a formula proposed by Blackwell and MacQueen to generalize Polya urn models (Skyrms 1996a, 326). It extends the γ-λ-system—on the grounds, however, of the purely subjective Bayesian reading of the former—to the case, where the outcomes take on real numbers in the unit interval. The γ-λ-system is inferred on the purely subjective Bayesian reading by adopting the following conditions: (1) exchangeability and non-independence; (2) sufficientness; (3) $t \geq 3$; and (4) strict coherence. In order to obtain a generalization (proposition 1.42) conditions (2) and (4) need to be modified accordingly:[64]

(2*) For any measurable set B_i and any finite sequence of observations:

$$P(X_{n+1} \in B_i \mid \mathbf{n}) = f_i(n_i);$$

(4*) For any measurable set B, which has a non-zero Lebesgue measure and for any finite sequence of observations n:

$$P(X_{n+1} \in B \mid \mathbf{n}) > 0.$$

Skyrms also brings out an alternative route to solve the problem of the value continuum, which corresponds to the objective Bayesian reading of the γ-λ-system. On this reading the system in question is equivalent to multinomial sampling with prior Dirichlet distribution. While there is a natural generalization of a Dirichlet distribution in the form of a Ferguson distribution,[65] the solution based on the latter is equivalent to the one discussed in the foregoing.

Admittedly, there is a serious drawback to using the Dirichlet priors (or Ferguson distribution):

> One would like to have a prior distribution for P with the property that if X is a sample from P and $X=x$, then the posterior guess at P gives more weight to values close to x than the prior guess at P does. For the Dirichlet process prior, the posterior guess at P gives more weight to the point x itself, but it treats all other points equally. In particular, the posterior guess at P actually gives less weight to points near x but not equal to x (Ferguson 1974, 622). I quote from (Skyrms 1996a, 330).

This problem was also recognized by Carnap, who called it the problem of *analogy by similarity*.[66] The fact that the sample gives more weight not only to *x*, but also close to *x*, would violate Johnson's sufficientness postulate. If so, then it is not possible to represent analogy by similarity within Carnap's systems.[67] However, Skyrms proposes a natural generalization of Carnap's system in order to enable the representation of analogy by similarity.[68] According to Skyrms the most natural generalization consists in finite mixtures of inductive methods that belong to the γ-λ-system. His proposal is to construe *hyper-Carnapian* methods by imposing *hyperprior* distribution on a finite number of meta-hypotheses concerning the values of the parameters γ_i and λ. The predictive probabilities of the γ-λ-system thus become conditional on the meta-hypotheses, which are themselves updated using Bayes's theorem.

There is also another kind of analogy recognized by Carnap, namely *analogy by proximity* in time or space.[69] A solution to this problem has been suggested by Kuipers (Kuipers 1988; Skyrms 1991). He points out that Carnap's inductive logic is already suitable for sampling from an urn with replacement. It is natural then that Carnap's inductive logic be applied to the urn model of a Markov chain, assuming, however, that transitions from a given state give us no information about transitions from a different state. It is inference about Markov chains that opens up the way to account for temporal patterns.

Carnap's suggestion was rather that we need to introduce into the language systems two-place predicates representing time-ordering. As the work of Tuomela and Hilpinen suggests, however, such an approach to this problem does not seem to yield unique solutions, unless to some extent arbitrary conventions are allowed (Tuomela 1966; Hilpinen 1966). In the solution proposed by Kuipers and worked out in detail by B. Skyrms, and S. Zabell the task of representing the temporal order is carried out, instead, by the indices (Skyrms 1991, 447).

As Zabell and Skyrms demonstrate,[70] the subjective condition which leads to Carnapian inductive logic for Markov chains has the form of Johnson's sufficientness postulate (Skyrms 1991, 452): the probability $P(S_j \mid S_i)$ of a transition to the state S_j, given the present state is S_i, depends only on i, j, the observed number $n[S_j, S_i]$ of transitions from S_i to S_j and $\Sigma_m n[S_m, S_i]$:

$$P(S_j \mid S_i) = (\lambda\gamma_{ij} + n[S_j, S_i]) \,/\, (\lambda k + \Sigma_{m=1}^{k} n[S_m, S_i]).$$

One of the most debated issues in the Carnap-Hintikka style systems of inductive logic is the problem of the choice of a particular inductive method from the continuum, i.e. the problem of assigning numerical value to the systems' parameters (e.g. λ or α). The problem was investigated in detail by Festa.[71] His approach is grounded on the objective Bayesian reading of the λ-γ-system of inductive logic, where it is presupposed that there are multinomial contexts with corresponding objective probability distributions. The latter, however, are assumed to be unknown in advance of performing experiments, and the uncertainty is reflected in the prior Dirichlet distribution, which is characterized by the following probability density function:

$$Dir(q_1,\ldots,q_{k-1};\alpha_1,\ldots,\alpha_k) = \frac{\Gamma(\alpha)}{\Gamma(\alpha_1)\ldots\Gamma(\alpha_k)} q_1^{\alpha_1-1}\ldots q_k^{\alpha_k-1}; \qquad \text{(DIR)}$$

where $\alpha_1,\ldots,\alpha_k > 0$, $\alpha = \Sigma\alpha_i$, and $q_k = 1-q_1-\ldots-q_{k-1}$ (Festa 1993, 61-62).

The important consequence is that the γ-λ- system and the class of prior Dirichlet Dir $(\alpha_1,\ldots,\alpha_k)$ on **q** are equivalent; where $\lambda = \gamma = \Sigma\alpha_i$ and $\gamma_i = \alpha_i/\alpha$.[72]

In order to determine a particular inductive method it is necessary to determine the values of the parameters (it is sufficient to determine γ and λ). With this aim in view Festa advocates an account which addresses some features of the context of scientific investigation.[73] In particular, he refers to the background knowledge of scientists in the specific form of "the corpus of theories and other assumptions taken for granted, at least within that inquiry", and also to "the cognitive goals pursued by scientists *qua* scientists" (Festa 1995, 300). The former is assumed to have a high degree of verisimilitude:

> A possible integration of these two ideas is given by the following assumptions: (1) the efficiency of an inductive method in approaching the truth depends on the degree of (dis)order of the sector of the universe under investigation; (2) hence, the adoption of a given inductive method as a presumably efficient tool to approach the truth can be based on appropriate presuppositions about the degree of (dis)order of the investigated sector of the universe (Festa 1995, 305).

In order to precisely express and measure the degree of (dis)order in the universe Festa employs *Gini diversity* $G(\mathbf{q}) = 1-\Sigma\mathbf{q}$; while the presuppositions with respect to the degree of (dis)order are taken as prior estimates of $G(\mathbf{q})$. If we suppose that the true value of $\mathbf{q}$ is q, then *the distance from the truth* is defined by Festa (Festa 1995, 306) as the distance between q and $E_\lambda(\mathbf{q}\,|\,e_n)$, i.e. the expected value of $\mathbf{q}$ calculated on the basis of the $\gamma°$-λ-system (with $\gamma°$ being the optimum γ-parameter):

$$D_2(E_\lambda(\mathbf{q}\,|\,e_n), q) = \Sigma[E_\lambda(\mathbf{q}_\mathbf{i}\,|\,e_n) - q_i]^2.$$

On these grounds Festa specifies the way to arrive at the optimum values of γ and λ. Suppose that the background knowledge contains information on a process similar to the one under investigation, and the values of $\mathbf{q}_\mathbf{i}$ specified thus are $\alpha_i^1,\ldots,\alpha_i^m$. Hence, the estimation of $\mathbf{q}_\mathbf{i}$ is as follows:[74]

$$Est(\mathbf{q}_\mathbf{i}) = (\alpha_i^1 + \ldots + \alpha_i^m) \,/\, m.$$

And this result via the fact that the marginal mean $E(\mathbf{q}_\mathbf{i}) = \gamma_i$ leads to the determination of the optimum parameter γ for all k categories (Festa 1993, 30; 98; 104).

Once the optimum γ-parameter is set, the optimum λ-parameter is also available. There are two independent ways (Festa 1995, 311-17) that lead Festa to establish the following optimum value of λ:

$$\lambda° = \frac{Est[G(\mathbf{q})]}{G(\gamma°) - Est[G(\mathbf{q})]}$$

Even if we adopt Festa's auxiliary assumptions, which allow us to establish the optimum inductive method within the γ-λ- system, it is still doubtful that the proposed solution is available on Carnap's own grounds. I postpone, however, the epistemological discussion to the next chapter.

Thus far I have outlined how the reading of the systems of inductive logic within a Bayesian framework contributes to the solution of difficulties, which were insurmountable within Carnap's initial framework: the problem of the value continuum, the problem of analogy, the representation of time-ordering and the problem of the choice of the optimum inductive method. There is still another gross difficulty that Carnap was not able to solve, i.e. the problem of zero-confirmation of non-tautological generalizations. This problem has already been discussed in greater detail with respect to Hintikka's modification of Carnap's systems of inductive logic (see page 31). The subjective Bayesian reading of the systems of inductive logic comes up, however, with a solution to the problem, which—in contrast to the one advanced by the Finnish School—is confined to a minimal modification of the γ-λ-system:

> In Hintikka's approach the predictive probabilities can depend on the value of T, the number of instantiated species; and the result is a mixture of continua on all possible subsimplexes of the support of the mixing measure in the de Finetti representation.
>
> The point of the present note is that a much simpler approach, meeting precisely the objection raised at the time Carnap published his monograph, is also possible (Zabell 1997, 282).

Given the γ-λ-system of inductive logic is generated by the sufficientness postulate, a slightly modified version of the postulate—as Zabell demonstrates—gives rise to a new system which avoids the difficulty in question:

> *the new sufficientness postulate*: for all $n \geq 1$, and all frequency vectors $\mathbf{n}$, the predictive probabilities $P(X_{n+1} = c_i \mid \mathbf{n})$ are functions of $f_i(n_i, n)$ of n_i and n alone, *except* when $n_i = 0$ and $T = 1$(Zabell 1997, 268);

where T is the number of categories observed in the sample.

Notwithstanding the modification, the resulting system is a member of the γ-λ-system unless all the outcomes observed in the sample fall within *one* category. In the latter case though, the respective generalization is positively confirmed by the evidence (Zabell 1997, 269).

One of the serious threats that Carnap was not able to overcome concerns the dependence of the confirmation value on the number of Q-predicates, or, in Bayesian terminology, on the number of categories. There is, however, a natural generalization of the systems of inductive logic, which leads to independence of the number of categories known in advance with respect to an experiment. J. Pitman introduced a three-parameter system, which seems to be a natural generalization of the systems of inductive logic, on the one hand, while on the other, it avoids the zero-confirmation problem as well as the problem discussed in this paragraph.[75] The recent emergence of this system constitutes the best response to all those proclaiming the Carnap-Hintikka program to be a dead end.[76]

Notes

1 Carnap's inductive logic was preceded by an alternative, and rather unexplored, approach more closely perhaps matching the pattern of the systems of deductive logic (Łukasiewicz 1913/1970); see also (Woleński 1997, 35-38).

2 For a survey of the 20th century attempts to provide a logical foundation for probability, and inductive inference, see (Hailperin 1990); (Howson 1995; 1993, 1-3).

3 Cf. e.g. (Chisholm 1966/1977, 38-39); (Hintikka 1980); (Kakkuri-Knuuttila and Knuuttila 1990); (Niiniluoto 1994/1995); (Kakkuri-Knuuttila 1997).

4 Cf. (Nagel 1955, 346-59); (Essler 1970, 13-57); (Milton 1987, section 2).

5 Cf. (Łakoma 1992, 50); (Hacking 1975, 87-88). It appears that the first attempts to use probability as a measure of the degrees of epistemic uncertainty could be traced back in *Logic* of Port Royal of 1662 (Chapter IV). These are then the first attempts to combine induction and probability (Hacking 1975, 76-77).

6 For an analysis of the metaphysical motivation see (Hacking 1967, 603-605).

7 However, R. Barcan Marcus argues that it is the epistemological analysis, of the kind I pursue throughout this book, that is adequate to analyze Carnap's modal concepts; cf. (Barcan Marcus 1981, 289).

8 Thomas Bayes (1702-1761) was an English clergyman credited with "Bayes's theorem", which, however, he never explicitly formulated; cf. (Bayes 1763/1958).

9 (Hume 1748/1963); cf. (Zabell 1989, 292). This problem was recently revived by K. Popper and D. Miller in (Popper and Miller 1983); for subsequent discussion see e.g. (Jeffrey 1984); (Redhead 1985); (Rodriguez 1987); (Miller 1990); (Mura 1990); (Howson 1991b); (Gemes 1997); (Kuipers 2000).

10 Cf. (Ramsey 1931/1964, 85-86) for a severe criticism.

11 For an outline of the problems of the classical definition of probability see (Earman and Salmon 1992).

12 For a discussion of the viability of this interpretation of probability see (Nagel 1945); (Reichenbach 1945); (Williams 1945); (Nagel 1955); (Galavotti 1995/1996, 100-104).

13 Cf. (Hacking 1975, 134). H. Jeffreys was in favor of the logical interpretation of probability, and therefore he was a remarkable exception among working statisticians; cf. (Kyburg 1966, 254).

14 Cf. (Keynes 1921/1948). For a criticism of his views see (Ramsey 1922/1989); (Ramsey 1931/1964); (Howson 1997b, 517-18). For a response see (Watt 1989).

15 Keynes formulates his argument in Chapter 3 of his *Treatise on Probability* (Keynes1921/1948). Keynes's idea of a comparative theory of probability was subsequently developed by B. O. Koopman in (Koopman 1940), T. Fine in (Fine 1973), and R. Wójcicki in (Wójcicki 1976).

16 In his early writings on probability, however, Carnap—following J. Hosiasson—referred to S. Mazurkiewicz's axioms rather than Kolmogorov's; cf. (Mazurkiewicz 1932). For a penetrating discussion of Kolmogorov's axioms see (Fine 1973, 58-84). In particular, T. Fine assents to the criticism leveled against the latter by Carnap, who claims that this

set of axioms is incomplete and inadequate because of "the supposedly close relation to the relative-frequency interpretation of probability" (Fine 1973, 59).

17 For a survey of other attempts at axiomatization of probability theory see (Dumitriu 1977, 105-9).

18 Cf. (Carnap 1930/1959, 134-41); (Essler 1970, 57-64).

19 Cf. (Tarski 1933/1956; 1944/1949). For a detailed discussion of Tarski's definition of truth based on the assumption of there being an intended interpretation of a given language—the assumption along Carnap's own view of semantics presented here—see (Wójcicki 1995/1996b). For a thorough analysis of Carnap's semantics underlying his systems of inductive logic see (Smokler 1977); (Hintikka 1981, 63-64).

20 The list of Carnap's writings on inductive logic to follow is far from complete, but it contains Carnap's most important works, especially from the perspective of the central argument carried out in this book: (Carnap 1945; 1945/1970; 1947; 1947/1997a; 1947/1997b; 1950; 1951a; 1951b; 1952; 1953a; 1953b; 1953c; 1953d; 1955; 1962; 1963b; 1963c; 1963d; 1968a; 1968b; 1968c; 1971a; 1971c; 1973; 1977; 1980); (Carnap and Stegmüller 1959). For a critical survey of Carnap's developments of inductive logic see (Carnap 1947/1997a; 1955); (Williams 1952, 104-109); (Shimony 1955b); (Nagel 1963); (Walk 1966); (Salmon 1967b, 68-79); (Pietarinen 1970, 144-46); (Carnap 1971b, 3-4); (Batens 1975, 70-104); (Jeffrey 1975, xxli); (Shimony 1975); (Stegmüller 1975, lxi); (Niiniluoto 1976, 93-96); (Bogdan 1980); (Jeffrey 1980a); (Rosenkrantz 1981, I.3); (Nekrasas 1987/1992); (Shimony 1992, 267-72); (Dubucs 1993a, 80-82); (Niiniluoto 1997, 117-20); (Kuipers 2000, 74-78).

21 (Carnap 1963a, 63). For a critical assessment see (Koj 1990, 278-81), and for a generalization of this concept, which allows for inconsistent and incomplete state-descriptions see (Voishvillo 1982).

22 For the argument that the concept of the degree of confirmation *is* an epistemic concept see (Sellars 1947, 647; 1964, 219); (Coffa 1974, 148-50). For a study of the general formal properties of confirmation see (Hempel 1943; 1945/1965).

23 In this respect Carnap followed the tradition of the "Cambridge School", most notably Keynes and Jeffreys; cf. (Carnap 1945/1970, 451); (Hempel 1945/1965, 21-25); (Creath 1998, 212-13). For a severe criticism see (Salmon 1965a; 1967a, 729-32; 1969); (Popper 1968, 132-33); (Niiniluoto 1976, 111-112).

24 It might be objected that it does not make much sense to present Carnap's initial logical frameworks, in which he presented his systems, for, on the one hand, they are subsumed by subsequent developments (cf. Kuipers 1978b), and on the other hand, the kind of logical framework employed is not relevant to the presentation of Carnap's system. The discussion of this argument I relegate to the next chapter, where, hopefully, the reasons for presenting also Carnap's initial frameworks, and subsequent modifications thereof, will become apparent.

25 The rationale for this procedure is elucidated by Carnap in (Carnap 1975).

26 Kemeny formulates there, however, a modification in the construction of state-descriptions: first, state the axioms and the intersection of their ranges, and then set up the measure function, cf. (Carnap 1950, 40). This modification proved crucial in later developments of the systems by J. Hintikka, I. Niiniluoto and T.A.F. Kuipers; cf. (Hintikka and Niiniluoto 1976/1980); (Niiniluoto 1977); (Kuipers 1978b).

27 This assumption of independence was successively alleviated by Carnap, for the first time in (Carnap 1952a). F. Schick claims that according to Carnap primitive predicates denote

psychologically simple properties; cf. (Schick 1967, 187). This claim, however, is not supported by Carnap's writings.

28 For an extended explanation see (Carnap 1950, 73).

29 W. C. Salmon was first to point out that although this condition is satisfied within each language system in Carnap's construal, it is nonetheless violated when we shift between different language systems, i.e. two synonymous sentences get different confirmation values in two language systems, which differ only with respect to the number of primitive predicates; for details, see (Salmon 1961, 256).

30 In fact, conventions are already embedded in Kolmogorov axioms: "the probability scale, axiomatized by Kolmogorov, is occasionally empirically meaningless and always embodies an arbitrary choice or convention" (Fine 1973, 59).

31 For an argument against basing a system of inductive logic on the standard probability axioms see (Cohen 1973); (Cohen 1980, 165-69).

32 Suppose for the sake of simplicity that $e \equiv t$. Let P be the only predicate of a given language system L_{11}, and j be the conjunction "$P(a_1) \wedge \ldots \wedge P(a_{10})$", while h be "$P(a_{11})$". As $h \wedge j$ constitute only one state-description in L_{11}, by the application of the classical definition of probability we obtain that $m(h \wedge j) = 1/n$ (where n is the total number of state-descriptions in L). Now, j holds in two state-descriptions: $h \wedge j$ and $\neg h \wedge j$, hence $m(j) = 2/n$; similarly, $m(h) = (n/2) / n = 1/2$. On this account then, $c(h|j) = 1/2 = m(h)$.

33 It is a violation of the condition of strict coherence then introduced by A. Shimony; cf. (Shimony 1955a). See also page 82.

34 See also (Skyrms 1991, 440).

35 For more details see (Carnap 1952b); (Hesse 1968a).

36 The same claim is advanced in (Salmon 1966, 17-18); (Bohnert 1975b, xlii); (Fodor and Lepore 1992, Chapter 2). For counter-arguments see (Smokler 1967); (Hacking 1969).

37 For important contributions see (Martin 1958); (Scott and Krauss 1966); (Kuipers 1978b); (Scott 1979).

38 (Kemeny 1963, 717-12); (Kemeny 1952, 38-42). He was also able to extend Carnap's system to predicate calculi of all finite orders, cf. (Kemeny 1953c); see also (Kyburg 1964, 21).

39 A similar view on generalizations is advanced by L. J. Savage, who claims that generalizations are "elliptical ... statements of a variety of finite conjunctions" (Savage 1967, 607).

40 See e.g. (Hintikka 1965b, 274-75; 1987a, 302-303; 1987b, 12-33).

41 For an extensive discussion on whether zero-confirmation constitutes a real problem see (Barker 1957, 87-88); (Michalos 1971b, 43-48). (Hintikka 1975a); (Essler 1975); (Kuipers 1976b, 200); (Creath 1998, 213); (Kuipers 2000, 84-87). On objections similar to and preceding Carnap's see (Zabell 1989, 306-308).

42 See also (Hintikka 1968b, 193-94; 1968c, 225); (Kuipers 1978a).

43 He developed the basics in (Hintikka 1953); for further developments see (Hintikka 1965a); (Hintikka and Rantala 1975, 42-50); (Scott 1979); (Hintikka 1987a, 277-80; 1987b, 14-16); (Rantala 1987).

44 They are supposed to describe "possible kinds of individuals"; cf. (Hintikka 1965a, 55); (Rantala 1987).

45 See also (Hintikka 1965a, 55).

46 For detailed derivations see (Hintikka 1966, 115-120); (Hilpinen 1968, 52-69).

47 In (Kawalec 1998) I analyze in more detail various attempts at an adequate interpretation of α.

48 One of the most compelling problems discussed in this context is the problem of the so called "transcendental hypotheses", i.e. hypotheses concerning the relations of the observed objects to unobserved; cf. (Hilpinen 1971, 273).

49 See (Jeffrey 1983, 133-45) for an explanation of what precisely constitutes Bayesianism. J. Hintikka points out that a change of the value of the parameters of the systems is not a Bayesian conditionalization; cf. (Hintikka 1968a, 304). A different objection against conditionalization on the grounds of the Carnap-Hintikka style systems was pointed out by Jeffrey in his letter of June 10th, 1957 to Carnap (Jeffrey and Carnap 1957/1958).

50 For an elaboration see (Stegmüller 1973, 418-502).

51 In fact, the first attempt to present an axiomatized system of inductive logic is presented as an appendix to (Carnap and Stegmüller 1959); see also (Carnap 1963c, 973-79); (Kemeny 1963).

52 More precisely, they present an axiomatic system which is adequate to represent positive confirmation values for generalizations, and the α-λ-system is a special case of the axiomatic system, cf. (Hintikka and Niiniluoto 1976).

53 It is also worth mentioning that Carnap's initial systems deviate from the standard probability theory in imposing measure over sentences (or propositions) rather than sets. For details see (Kuipers 1978b; 1980). I discuss the issue in more detail in (Kawalec 2000a); see also (Howson and Urbach 1993, 31-32); (Humphreys 1995/1996). In general, it appears that there are two different approaches to explicate the concept of confirmation: it is either introduced by means of explicit definition, or is implicitly defined by means of an axiomatized theory of probability followed by an interpretation or model for that theory; cf. (Humphreys 1998, 702).

54 Cf. (Kuipers 1978b, 40-43).

55 For the comparative chart see (Kuipers 1978b, 107).

56 It was definitely R. C. Jeffrey who first proposed the rudiments of the Bayesian reading of Carnap's inductive logic in (Jeffrey 1957) and who influenced Carnap to change his views; cf. (Carnap 1962). In the present study I do not account, however, for the conceptual shift in Carnap's thought from 1962 onwards (Carnap 1962; 1963c; 1971a; 1971b; 1980), for it introduces a drastic epistemological revision of Carnap's earlier assumptions. In this I follow the standard view of Carnap advanced first by W. Stegmüller; cf. (Stegmüller 1973); (Hilpinen 1973). R. J. Bogdan claims that Carnap's shift is the best evidence of the failure of the preceding program in inductive logic; cf. (Bogdan 1980, 406-9). Carnap perceives this shift as a gradual development of his earlier views, which is primarily due to his cooperation with R. C. Jeffrey; cf. (Carnap and Jeffrey 1971a).

57 For the concept of exchangeability see (De Finetti 1937/1964, 118-47); (Hintikka 1971). In Carnap's original system the role of the concept of exchangeability is played by that of symmetry: the permutation of individual constants does not affect the probability, cf. also (Skyrms 1991, 453).

58 This is proved in (Zabell 1982); see also (Zabell 1997, 267-68).

59 My exposition of the multinomial contexts is due to R. Festa's definition thereof, cf. (Festa 1995, 301). For a similar account see (Rosenkrantz 1980, 79).

60 This view is put forward by Skyrms, cf. (Skyrms 1996a, 323). However, Festa gives a detailed defense of the claim that the prior distribution reflects some objective features of the cognitive context of scientist; (Festa 1993, 1995).

61 Cf. (Hesse 1980, 214-17).
62 See e.g. (Lenz 1956).
63 In addition, it is further required that the measure be absolutely continuous with respect to the Lebesgue measure of the unit interval, cf. (Skyrms 1996a, 326).
64 For a sketch of a proof see (Skyrms 1993c, 85-86; 1996, 329-30).
65 For more details see (Skyrms 1993c, 84); (Skyrms 1996a, 327).
66 Cf. (Carnap 1980, 32-83); (Kuipers 1984b; 1988; 2000, 82-84); (Niiniluoto 1980, 228-34; 1981a; 1988a); (Spohn 1981).
67 For a demonstration see (Hesse 1964, 320-22). M. Hesse correctly points out that this fact is due to the symmetry conditions embraced by Carnap; cf. (Hesse 1968b); (Festa 1997, 243-46).
68 See (Skyrms 1996a, 330). This approach is further developed by R. Festa (Festa 1997), who develops mixtures of Carnapian methods for analogy by similarity for predictive inferences. Moreover, Festa thereby obtains exchangeable analogical methods; cf. (Festa 1997, 243-46).
69 The fact that Carnap was not able to demonstrate that his systems of inductive logic are capable of representing time-ordering invoked Putnam's sharp criticism; see (Putnam 1963); (Bar-Hillel 1968a, 21).
70 See (Zabell 1995, 175-78); (Skyrms 1991, 453-56; 1996, 333).
71 Cf. (Festa 1993; 1995). For similar attempts see (Lewis 1971); (Pietarinen 1974); (Kuipers 1984a); (Essler 1986); (Kuipers 1986).
72 For the proof see (Festa 1993, 65-66).
73 The first to suggest this kind of approach was D. Lewis in (Lewis 1971); a similar approach is also advanced in (Essler 1986). Festa follows here the so called *presupposition* view of induction. The most eminent advocate of this position was J. S. Mill, who insisted on the principle of uniformity of nature as indispensable for carrying out inductive inferences; cf. (Rosenkrantz 1981, Ch. 1.3, p. 5). With respect to the analyzed systems of inductive logic this position was advocated by Hintikka (Hintikka 1966) and later on by Pietarinen in (Pietarinen, 1972, 111); see also (Burks 1953); (Uchii 1972; 1973); (Batens 1975, 285-95); (Hintikka 1987, 303-304); (Gemes 1997); (Hintikka 1997, 318). However, Carnap strongly opposed this view, and always emphasized the fact that the theorems of inductive logic, i.e. sentences of the form $c(h \mid e)=r$ are *analytic*.
74 A further sophistication, which allows for the comparison of the similarity between the two experiments under question, is possible, cf. (Festa 1993, 104).
75 For more details, see (Zabell 1996). S. L. Zabell argues that already Carnap appreciated the role of studying the case of inductive inference when the number of categories is not fixed in advance; cf. (Zabell 1992, 229-30).
76 This is in direct opposition to the claims propounded by Carnap's closest allies; cf. (Bar-Hillel 1975; xxx), and to recent views of Hintikka; cf. (Niiniluoto 1997, 120-129).

Chapter 2

INDUCTIVE LOGIC AS A STRUCTURAL RELIABILIST THEORY OF JUSTIFICATION

For Carnap there is no other way to contrive a theory of the concept of inferential justification other than to construct the systems of deductive and inductive logic is . It is not, however, the kind of theory that would be propounded in the mainstream epistemology. In order then to motivate the discussion of Carnap's inductive logic as constituting a *theory of justification* I need first to introduce the epistemological framework appropriate for this discussion. I convey the difference between the kind of the theory of epistemic justification brought out by the Carnap-Hintikka style systems of inductive logic and the kind of the theory of epistemic justification espoused in the mainstream epistemology in terms of *minimalist* and *mentalistic* kinds of the theory of epistemic justification.

Furthermore, Carnap's minimalist theory of inferential justification seems to represent a kind of reliabilism. In discussing process reliabilism I set out the framework for the kind of the minimalist reliabilist theory of justification, namely *structural reliabilism.* I then proceed to examine Carnap-Hintikka style systems of inductive logic as constituting a structural reliabilist theory of justification.

1. The Mentalistic Theory of Justification and Process Reliabilism

Apparently, the bottom-up strategy prevails on the way the mainstream epistemological theories of justification are developed. A sample pattern of actually upheld beliefs is chosen at the outset, and is then subjected to subsequent modification and enlargement. It is against this set that the adequacy of the epistemological position under scrutiny is assessed. The other important factor in the evaluation of the viability of the epistemological stance is the set of contested epistemological positions. The examined position is supposed to sustain the objections of the opposite camps, and only then can it regain its plausibility. This rough delineation of how the theory of justification is brought out in contemporary epistemology motivates what is apparently assumed to constitute a necessary condition of such a theory of justification. In the ensuing I will refer to a *mentalistic* theory of justification whenever it satisfies the following condition:

DEFINITION 2.2 *Mentalistic theory of justification* (METJ)

T *is* a mentalistic *theory of the epistemic concept of justification only if the term "justified" occurring in* T *is a relational term defined over a tuple* <S, $\mathbf{B}_t$>, *where S is an epistemic agent, and* $\mathbf{B}_t$ *is a set of beliefs actually upheld by S* (*where t ranges over a part of, or even entire, S's lifespan*), *and* $\mathbf{B}_t$ *is closed under the operation of S-computational consequence of* $\mathbf{B}_t$.

By saying that "*S* is an epistemic agent" I mean *S* to be a human being. It is difficult to spell out further constraints imposed on *S*, which will hold in general, i.e. for all kinds of epistemological analyses which may be broadly characterized as mentalistic . For instance, *S* may actually be a brain-in-a-vat, a person being deceived by a malicious demon, or an idiot savant (Sosa 1991, 3). Regardless of individual differences, epistemic agents usually are assumed to have normal cognitive capacities, e.g. are assumed to be capable of carrying out processes of perception, memorization, or reasoning. A more detailed specification of what precisely constitutes these "normal" cognitive capacities will inevitably require recourse to an ontological discussion. The latter, however, epistemological debates often strip away.

I choose *S* to be an individual rather than a group, for it seems psychologically more plausible[1] to account first for how and why individuals generate, sustain and change their beliefs, and only then to proceed to the more complex case of social beliefs.[2] The mentalistic character of a theory does not, however, turn on the priority of the individual case over the social case as assumed in the present version of the definition of mentalistic theory of justification. Hence, should there be good arguments against the assumed priority, definition 2.2 will demand an appropriate modification.

The concept of "*S*-computational consequence of *X*" reflects the following feature of epistemological discussions. Suppose that *S* believes *p*, and *S* believes $p \wedge q$. Notwithstanding, for the evaluation of the epistemic status of *S*'s belief in *p* it is important that *S actually* carries out the process of inferring her belief in p from her belief in $p \wedge q$. The logical consequence is not relevant for the evaluation at hand, unless S performs accordingly the inference herself.

Interestingly, this fact was clearly recognized by R. Descartes (see his *Reply to Objections II* in (Descartes 1967, 38-39)). R. Audi formulates this point succinctly thus:

> it is reasoning processes and not abstract structures that I call deductive or inductive. I do not take arguments, as abstract structures, to be intrinsically of either kind, though we speak of them as deductive or inductive so far as they seem best assessed by deductive or inductive standards (Audi 1998, 162).

Thus, if S believes in p on some other ground than her belief in $p \wedge q$, then, epistemologists assume, it is not relevant to the epistemic status of S's belief in p that p is in fact a logical consequence of $p \wedge q$.

A. I. Goldman presents two arguments to demonstrate that the logical consequence is not a sufficient condition for justification to be stated on the grounds of a mentalistic theory of justification. Suppose S realizes a logical consequence of her belief in $p \wedge q$, then S might reject $p \wedge q$ instead of justifiedly believing the consequence. The other argument is that even if S beliefs $p \wedge q$, but nonetheless fails to perform the inference herself, then S does not justifiedly believe in p (Goldman 1986, 83-84).

The logical consequence obtaining between S's beliefs is not a sufficient condition for justification either, because it will require logical omniscience on the part of epistemic agents, and this requirement is implausible in respect of actual epistemic agents. One reason is worded by G. Harman thus: "If you believe ... premises and then deduce [the] conclusion, you learn that your beliefs imply that conclusion. That by itself is not enough for you to be able to be justified in inferring that conclusion. It may instead provide a reason for you to question your belief in the premises" (Harman 1993, 204).[3] This is also why S is not taken to be epistemologically responsible for not being able to track all logical relations among her beliefs. For that reason I use in my definition of mentalistic theory of justification (definition 2.2) the concept of "S-computational consequence" meaning thereby the subset of logical consequences among S's beliefs that S, due to her confined cognitive capacities, is able to track herself at a given time. "S-computational consequence", therefore, results in constraining the set of possible reasons or grounds that might serve in justifying S's beliefs, and limits S's epistemic responsibility.

A. C. Grayling coined the term "finitary predicament" to refer to the limitations of empirical cognitive agents: "It is an all too familiar fact that the epistemic capacities of humans are finite. We suffer the finitary predicament; our empirical resources for acquiring and testing information about the world are limited, and so are our powers of inference, analysis, recognition and memory" (Grayling 1995, 10-11). This, however, is only possible when psychological or even neurophysiological concerns underlying *S*'s cognitive capabilities are taken into account. There are, thus, important descriptive factors included in definition 2.2.[4]

This definition stripes away other constraints that would result in further diversification among the positions on mentalistic theories of justification.[5] One of the constraints concerns what kind of epistemic access *S* has to her beliefs might be epistemologically relevant; the two broad positions being here internalism[6] and externalism.[7] The other constraint imposes a specific ordering on the belief set $\mathbf{B}_t$ or on the set of justified beliefs $\mathbf{JB}_t$; there being, again, two broad categories: foundationalism[8] and coherentism.[9]

(METJ) does not specify the set of possible reasons or grounds for *S*'s beliefs. There is no consensus among epistemologists as to what might take on this role; for instance some would agree that *S*'s first-order beliefs count as relevant, some would add that also *S*'s higher-order beliefs are relevant, while others will deny the latter claim or both.

However, no particular definition of belief has been presupposed by (METJ); in fact, this definition might serve as a contextual definition of the concept of belief. For on the one hand, we have already observed that *S*'s computational capacities play an important role with respect to her beliefs. That, however, would require that her beliefs be, at least partly, amendable to *S*'s computational capacities, as they must be upheld by her. Moreover, (METJ) implies that beliefs can be sustained by actual epistemic agents, which is tantamount to acknowledging that they are kinds of propositional attitudes. This is the standard view, which—as J. Heil admits—"has proved remarkably resilient" (Heil 1993, 48). In general, propositional attitudes are mental states with propositional content. Beliefs are the kind of propositional attitudes which involves "a stance on the question whether the proposition is true, quite apart from whether it would be nice if it were true" (Goldman 1986, 14).

On the other hand, we learn from (METJ) that it is primarily beliefs that we confer justification upon, i.e. beliefs can be rightfully evaluated epistemologically. They are a kind of, and not just *the*, as argued below, justification-bearers.[10]

In order to have some flavor of how (METJ) works let me apply it to one of the epistemological positions, to wit, Goldman's position developed in "What Is Justified Belief?" (Goldman 1979/1993).[11] Goldman's theory of justification is constructed on the basis of the following definitions:

(BC) If *S*'s belief in *p* at *t* results from a reliable cognitive process, and there is no reliable or conditionally reliable process available to *S* which, had it been used by S in addition to the process actually used, would have resulted in S's not believing p at t, then S's belief in p at t is justified (Goldman 1979/1993, 102).

(RC) If S's belief in p at t results ("immediately") from a belief-dependent process that is (at least) conditionally reliable, and if the beliefs (if any) on which this process operates in producing S's belief in p at t are themselves justified, then S's belief in p at t is justified (Goldman 1979/1993, 98).

Condition (BC) is called by Goldman "the base condition" for it concerns the kind of belief-forming processes, which do not take beliefs as inputs.[12] The recursive clause (RC) states the reliability condition for those belief-forming processes, which take beliefs as input. Thus, the reliability of the latter kind of belief-forming processes is conditional on the reliability of input beliefs. For instance, my visual belief "There is a sheep in front of me" formed while perceiving a sheep from a distance of two meters in normal conditions (including the correct functioning of my perceptual and cognitive apparatus) would be an example of a belief which satisfies (BC). While my belief recalled from memory that there was a sheep in front of me on September 21st, 1999 is reliable, unless the belief in question was not justified.

Both (BC) and (RC) apply the phrase "S's belief in p... is justified", therefore prima facie satisfying definition 2.2 as regards the domain of the concept of justification. What we still need to resolve is whether Goldman's understanding of "S" and "$\mathbf{B}_t$" satisfies it as well. In the case of "S" it holds trivially, since Goldman asserts that S is a person (Goldman 1979/1993, 92). From (BC) and (RC) we learn that the members of the domain $\mathbf{B}_t$ of justification can serve both as inputs and outputs of the psychological processes of a given individual.[13] The closure condition (METJ) imposes on $\mathbf{B}_t$ is, therefore, satisfied in Goldman's account. For it is not only that S-computational consequences belong to $\mathbf{B}_t$, but also all sorts of outputs of psychological processes, which take beliefs as input, for instance wishful thinking or hasty generalization.

2. Conflicting Meta-Level Assumptions of Mentalistic Reliabilist Theories of Justification

Apparently, Goldman advances a mentalistic theory of justification. In order to give it some substance he brings out some other meta-level features underlying his process-reliabilist theory of justification. One of Goldman's motivations, which leads him to a mentalistic theory of justification, is that his theory must be "quietistic" with respect to "our ordinary standards". A theory is *quietistic* if it minimizes the prescriptive intervention in a given discourse:[14]

DEFINITION 2.3 *Quietistic theory* (QT)

Theory T of a given concept C is quietistic *with respect to a given discourse D iff the standards that T prescribes for a usage of C within D do not differ and improve upon the standards of D.*

It would appear that it is only *D*, or, generally, one of its improper subsets, that could constitute a reference-class against which to assess the quietistic character of a given theory of a concept *C*.

The quietistic character may or may not conflict with another important feature of Goldman's theory:

DEFINITION 2.4 *Substantive theory* (ST)

Theory T *of the concept of justification* J *is* substantive *iff the necessary and sufficient conditions* T *lays down for* J *are stated in non-epistemic terms.*

Goldman acknowledges that given (QT) it is impossible to escape the vagueness of ordinary standards with respect to what counts as an epistemic concept.[15] Nonetheless, among the clearly epistemic concepts he lists the following: "'justified', 'warranted', 'has (good) grounds', 'has reason (to believe)', 'knows that', 'sees that', 'apprehends that', 'is probable' (in an epistemic or inductive sense), 'shows that', 'establishes that', and 'ascertains that'". Whereas the list of undoubtedly standard non-epistemic concepts[16] contains the following: "'believes that', 'is true', 'causes', 'it is necessary that', 'implies', 'is deducible from', and 'is probable' (either in the frequency sense or the propensity sense). In general, (purely) doxastic, metaphysical, modal, semantic, or syntactic expressions are not epistemic" (Goldman 1979/1993, 90).[17]

Before I proceed to investigate the role of (ST) in Goldman's theory, let me briefly examine the two lists of concepts. There is no point in trying to be far more explicit than Goldman on what constitutes the epistemic—non-epistemic dichotomy. Note, however, that one of the delineations separating the two is that it is, by and large, the former that conveys propositional attitudes. Moreover, it is not accidentally so, because what seems to be a necessary condition for any epistemic relation to obtain is that human psychological functioning be involved. This point can be illustrated on the example of "is probable". This expression figures in both Goldman's lists, but with clear indication in parentheses whether it is applicable to the epistemic agent's gradual beliefs, or to physical properties. Still, apart from human attitudes, all expressions on the former list imply the propositional content of these non-idealized attitudes including a that-clause.

However, once we agree that referring to non-idealized propositional attitudes is a necessary condition for a theory not to be substantive, then making (ST) compatible with (QT), seems highly problematic.[18] Consider the following example. Jones, asked why a belief in *p* is justified, answers thus: "I am inclined to think that a belief in *p* is justified, because I am assured by a most trustworthy expert economist that *p* is justified". Apparently, Jones's answer satisfies (BC) and (RC). The process of testimony is conditionally reliable in this case, for it very often turns out that whenever the expert economist tells Jones that *p* is justified, it is in fact true that *p* is justified. The testimony, however, is only conditionally reliable on the justification of input beliefs. Suppose, though, that the expert economist's belief that *p* is justified is a result of a reliable process and she does not have undermining evidence against it. Thus, (BC) is also satisfied. Therefore, the sentence uttered by Jones seems to fit Goldman's theory well. Nonetheless, it is incompatible with a conjunction of properties (ST) and (QT) that Goldman projects for his theory. Why is this so? Suppose we want to hold fast to (ST). Jones's sentence, however, which aptly expresses a part of his epistemic evaluative standards employs the epistemic term "justified". Hence, in order to stick with (ST) we need to discard "justified" in the sentence uttered by Jones by substituting "*q*" instead of "p is justified". What we obtain is a perfectly meaningful sentence: "I am inclined to think that a belief in *p* is justified, because I am assured by a most trustworthy expert economist that *q*". This sentence also satisfies (BC) and (RC) in a way similar to the initial sentence uttered by Jones. However, what Jones initially says then is tantamount to specifying the process, by means of which he is able to solve the problem of evaluating the justificatory status of the belief in *p* by reducing it to the problem of checking the expert economist's opinion whether *p* is justified:

$$\mathrm{EET}(Jp) \rightarrow JBp; \qquad \text{(i)}$$

where "EET(*p*)" stands for "the expert economist states that *p*"; "*Bp*" stands for "believes in *p*"; and "*Jp*" stands for "*p* is justified".

The new sentence obtain by substituting "*q*" in place of "p is justified" reads:

$$\mathrm{EET}(q) \rightarrow JBp. \qquad \text{(ii)}$$

In accordance with (ST), clause (i), in opposition to (ii), could not be satisfactory, since in stating part of Jones's evaluative standards it uses the evaluative term "justified", which is then discarded in (ii). Nonetheless, (ii) violates (QT). Jones's initial utterance establishes the relevance between what he takes into account, and what the expert economist tells him. They are both considering the same proposition *p*, and moreover they both assume that the sort of evaluation to be relevant. However, once we drop the epistemic term from Jones's initial statement of evaluative standards, as is done in (ii), then neither of the two kinds of relevance pointed out above could be preserved. So, it well accords with (ii) that Jones would take the belief in the proposition that "one of the European stock markets is going to collapse within the next two months" as justified, even though the expert economist told him that "none of the European stock markets is going to collapse within the next two months". It seems that (ii) is a misconstrual of Jones's evaluative standards for the justificatory status of beliefs, and thus (ii) violates (QT).

In order to make my point perspicuous I referred in the foregoing argument to testimony. Nevertheless, the very same point can be established for any kind of belief-forming process. Of course, this is a manifestation of the *Generality Problem,* i.e. the problem of how to specify in non-epistemic terms the relevant class of processes which determine the reliability of its member-processes, and to which belongs the process whose reliability is under scrutiny (Alston 1995b, 2).

In the initial sentence uttered by Jones "justified" from being a part of his epistemic evaluative standards becomes a part of the content specification of what the expert economist tells Jones. Thus, the role Jones assigns to "justified" in the expert economist testimony is not evaluative, Goldman might want to claim, but is that of identifying the relevant utterance of the expert economist. Therefore (ST) and (QT) seem perfectly compatible with regard to Goldman's theory explicating the epistemic evaluative standards of Jones. For suppose we now distinguish between theories quietistic with respect to ordinary standards of epistemic evaluation (QTE) and theories quietistic with respect to ordinary standards of language usage (QTL). So, (ST) and (QTE) are compatible within Goldman's theory, while it is (QTL) that is violated by the theory in question.

The strategy of defending the compatibility of (ST) and (QT) within Goldman's theory seems not promising, after all. For in Goldman's account the necessary condition for (ST) requires the identification of a proposition in question across the epistemic—non-epistemic dichotomy. He cannot then distinguish (QTE) and (QTL), and then proceed to reject only the latter on pain of assuming some mysterious access to evaluative standards of the justificatory status of beliefs which have propositional content. If we reject

(QTL), what kind of access to ordinary standards are we supposed to have such that these standards are (QTE)? (QTL) amounts to acknowledging that we cannot trust people's utterances, while (QTE) requires that we have access to the propositional content of their relevant attitudes. Even if we work out a way to cross identify propositions without reference to their content, it still will not assure that this cross-identification does not violate (QT), which it apparently does.

Another way to undermine the existence of the tension between (ST) and (QT) is by pointing out that Jones's utterance violates definition 2.2, since the expert economist is supposed to attribute "justified" to *p*, instead of *Bp*. Thus we need to reformulate Jones's initial utterance so that it would fit with definition of mentalistic theory of justification. Having accomplished this we could proceed to introduce—along the lines of (ii)—a reformulation of Jones's utterance to obtain an utterance compatible with definition 2.2, i.e. we could obtain it by substituting *q* for this part of Jones's evaluative standard, which contains epistemic concepts. This, however, would force us to replace "belief in *p* is justified" by *"q"*:

$$\mathrm{EET}(q) \rightarrow q. \tag{iii}$$

This seems even worse than (ii), since we have no reason at all to identify Jones's utterance as expressing his evaluative standards.

Still another way of undermining the incompatibility between (ST) and (QT) is to appraise negatively the relevance of Jones's utterance for his overall evaluative standards. Take it for granted that Jones expresses in his initial utterance one of his evaluative standards. Still, it has slight significance for Jones, who is used to applying all his epistemic evaluative standards but this one. Independently of how we estimate the plausibility of such a claim, it nevertheless violates (QT). Moreover, recall that the initial question concerns the evaluation of the justificatory status of *p*, and if the answer Jones gives us is the one that comes to his mind first, then we have no compelling reason to undermine the psychological plausibility of Jones's attributing at least some importance to the standard he utters.

Goldman might still want to undermine the relevance of Jones's utterance for epistemic evaluative standards, since he concerns his theory to be *nomic*:

(NT) A theory *T* of the concept of justification *J* is *nomic* only if *T*'s necessary and sufficient conditions apply to the set of justification-bearers of *J* without partitioning the set or with partitioning the set in such a way that the number of elements in each partition greatly outrun the total number of the partition-cells (Goldman 1979/1993, 90).

The idea here is that T should not be a collection of necessary and sufficient conditions expressed for each belief separately, but should be cast in general terms. And the level of generality should be as high as possible, and that is robustly manifested by the number of partition-cells; in particular Goldman's own theory imposes only a two-class partition. From this perspective, it might be argued, taking Jones's utterance as relevant for epistemic evaluative standards will force us to reject (NT), and deal with each belief separately, or, at least, introduce a very large number of types of beliefs. What has to be appreciated in this point is that Jones's evaluative standard of the justificatory status of beliefs is incompatible with Goldman's (BC) and (RC). Nevertheless, it is not (NT) that Jones's standard violates, but (ST). For suppose that Jones accepts Goldman's (BC) and (RC), and appends them with a further (Jones) base clause condition:

DEFINITION 2.5 *Base condition* (Jones)

If S's belief in p at t results from the expert economist's testimony to S that p is justified by a reliable cognitive process, and there is no reliable or conditionally reliable process available to S which, had it been used by S in addition to the process actually used, would have resulted in S's not believing p at t, then S's belief in p at t is justified.

The (Jones) base clause would not violate (NT) since it adds only one class to the initial Goldman's partition. However, it would inevitably violate (ST), for it is not only the fact that the base clause condition (Jones) uses the epistemic term "justified", but this use of epistemic terms would spread out in the whole theory via (RC).

Notwithstanding the case of Jones's utterance, (ST) and (QT) are even prima facie incompatible, because (ST), on Goldman's account at least, is clearly motivated by a reductionist strategy with respect to epistemic concepts, while (QT) is not.

Goldman seems to have intuitively tried to alleviate the tension between (ST) and (QT) persistent in his theory. The prevalent solution is to reformulate the necessary condition for (ST), so that it does not prevent epistemic concepts from playing a role in setting the epistemic evaluative standards—call it the alleged epistemic reformulation of (ST) "(ST-E)". But this solution is not available to Goldman.

Let me start my argument by appraising the role (ST) assumes in his theory. First, (ST) is essential for establishing the most plausible candidate for (BC) *within* Goldman's theory. Thus, all candidates which have prima facie plausibility, but which contain epistemic terms, are deemed to be appropriately reformulated, at best.

Second, Goldman insists on the *explanatory* character of his theory:

> I seek an explanatory theory, i.e. one that clarifies the underlying source of justificational status, it is not enough for a theory to state "correct" necessary and sufficient conditions. Its conditions must also be appropriately deep or relevant...[, and] it must be couched at a suitably deep, general, or abstract level (Goldman 1979/1993, 90).

The conditions (ST) and (NT) taken together seem to be sufficient and necessary to grant the explanatory status to a given theory. For, on the one hand, (ST) warrants that no epistemic terms are used, in particular, we avoid a merely synonymous substitution for the term "justified", or verbal definition thereof. On the other hand, (NT) grants that the way we apply the non-epistemic terms not only elucidates each particular case separately, but embraces them all at once.

Third, (ST) does much more for Goldman's theory than that; (ST) proves crucial in his rejection of other epistemological theories. As it seems, Goldman's strategy against candidates for (BC) stemming from other epistemological theories is essentially the following. Take a (BC) candidate, most of which use epistemic terms, and reject it according to (ST). Next, reformulate the candidate appropriately, so that it no longer contains epistemic terms. Finally, put it under scrutiny and examine whether it is a successful (BC) candidate. This sort of argumentative strategy though makes all these candidates doomed to failure from the outset. Why? Because they were brought out as plausible in their initial epistemic formulation rather than non-epistemic one. This is, at least, what the principle of charity would reasonably prescribe us to believe in this case. Thus (ST) is essential for the construction of Goldman's own theory insofar as it forces the possible (BC) candidates, which stem from alternative theories, to take a much less plausible position.

Having recognized the prominent role of (ST) within Goldman's theory we can return to the initial question, which concerns the applicability of (ST-E) to Goldman's theory. Given the above argument, it is clear that (ST-E) will lead to the undermining of the identity of Goldman's theory against the background of alternative theories. For (ST-E), in opposition to (ST), blocks the essential part of Goldman's argument against (BC) candidates, which stem from alternative epistemological theories.

In sum, while (ST) is in fact indispensable for Goldman's theory, there is no conceivable way how to appropriately modify it in order to accommodate (QT). Therefore, in the face of the conflict between (ST) and (QT) it is apparently the latter that needs to be given up within the theory in question. However, it would be very undesirable to reject (QT) *tout court*, for it embraces the role of the primary condition of adequacy for a mentalistic theory of justification. The rejection of (QT) would also seriously threaten Goldman's theory.

S. Cohen gives a different argument against (QT):

> the justification theorist is investigating the ordinary concept of justified belief—the concept that is expressed by our ordinary use of "justified belief" ... [W]e construe the justification theorist as trying to determine what we mean, in a very broad sense, when we ordinarily say of people that they have justified beliefs ... The problem with this way of viewing matters is that "justified" is a term we do not ordinarily apply to beliefs ... There is no ordinarily used expression "justified belief" (Cohen 1995, 116).

Once, however, we consent to dispense with (QT), Cohen's conclusion would be inevitable: "[t]here is no significant philosophical issue regarding justified belief" (Cohen 1995, 121). The project of minimalist theories of justification as brought out in this book need not, however, turn on this radical view.

While neither (ST) nor (QT) can be easily discarded, and together they are problematic, let me suggest another way that might prove helpful to delineate between these two conflicting meta-level properties of Goldman's theory. I cited Goldman's theory as a sample a mentalistic theory. So, let me now consider to what extent (METJ) might prove helpful to assess (ST) and (QT). Definition 2.2 does not overtly acknowledge what kinds of standards shall be used to determine an adequate condition of justification. Thus, it is seemingly neutral between (ST), which requires the standards to be stated in non-epistemic terms, and (QT), which requires the standards to preserve the ordinary evaluative standards. By the same token it would seem that (METJ) could not be read as opposing either (ST) or (QT).

However, this is only partially correct. (METJ) is neutral with respect to (ST), but seems less with regard to (QT). Recall that the concept of "*S*-computational consequence" was motivated by the quietistic character of Goldman's theory, in particular by the consideration that drawing upon "logical consequence" instead of "*S*-computational consequence" will be implausible in view of the actual epistemic agent's cognitive capacities. The latter, however, was introduced on the basis of the concept of belief, which is a propositional attitude, and which can be generated, sustained and changed by actual epistemic agents. Hence, dropping (QT) will also undermine the rationale for definition 2.2, and therefore if (QT) comes into conflict with another desirable property of a given epistemological theory, the definition of mentalistic theory of justification contains a rather straightforward indication that it is (QT) that shall be retained in some form or other.

One might also argue that (ST) is incompatible with (METJ), because (ST) cannot be appropriately defined for the same domain as (METJ).

To unravel this point let me recall Goldman's distinction between epistemic and non-epistemic terms. The term "is probable" was present in both groups, with a short comment by Goldman on the intended meaning. Thus "is probable" is an epistemic concept when it is not a frequency, chance, or propensity concept of probability; otherwise it belongs to the second group. It has been taken for granted that probability in the non-epistemic sense is ascribed to events, processes, occurrences, states of affairs or some other parts of the mind-independent world. And if it is applied to propositions or sentences, it is only derivatively so, because we can identify the objects we want to call "probable" without any reference to "propositional content", but by referring only to their physical or mathematical properties. Definition 2.2, on the other hand, does not endorse the existence of anything else apart from the propositional attitudes required as a necessary condition for a fully legitimate theory of justification. The discomfort (ST)-motivated epistemological theory finds with definition 2.2 is also recognized as the generality problem.[19]

We have, however, already noticed that while Goldman's theory cannot be sustained without (ST), the latter is incompatible with (QT). Now, it turns out that dropping (QT) will also require discarding (METJ). It looks as if Goldman's theory is caught between a rock and a hard place: either it is forced to retain (ST), and dispense with both (QT) and definition 2.2, or it retains (QT) and (METJ), and discards (ST). In either case though the identity of Goldman's theory may turn out to be spurious. This dilemma caused by internal tensions within the various requirements that motivate Goldman's theory is well taken and the moral is then that the theory cannot be sustained without giving up a substantial portion of it. Which one? Below I will suggest one alternative, which seems to restore the substantial motivation underlying (ST), and which also preserves the intuitions underlying (QT). The latter, however, has to be modified accordingly to reflect the fact that definition 2.2 is to be subsumed by a new requirement.

3. The Minimalist Theory of Justification and Structural Reliabilism

As argued in the foregoing, a plausible theory of justification should include a suitably reformulated (ST). A firm grasp of what (ST) was initially intended for will perhaps allow for delineation of its essential part, if any. Goldman articulates his primary intention thus:

> The term "justified", I presume, is an evaluative term, a term of appraisal. Any correct definition or synonym of it would also feature evaluative terms. I assume that

such definitions or synonyms might be given, but I am not interested in them. I want a set of *substantive* conditions that specify when a belief is justified (Goldman 1979/1993, 90).

He seems to be suggesting here that the theory is substantive if its account of the necessary and sufficient conditions of the justificatory status of belief is not limited to the mere acquaintance of epistemologists with the appropriate rules governing language usage. What seems crucial is that the theory should also provide us with the criteria of adequate evaluation of the justificatory status of *S*'s belief which have to be independent of the usage of epistemological language and the introspective capacities of *S* to reflect upon her own beliefs.

Apparently, Goldman's motivation underlying (ST) does not seem to be necessarily committed to anything like definition 2.2. Hence, it might be worth investigating the abandonment of mentalistic nature of the theory, keeping in mind, however, that we want also to stick to (QT). Is there any domain, which as belief, will seem to constitute natural epistemological kinds of justification-bearers? While the answer is so easily accessible, and was indeed, as already pointed out in section 1 of the first chapter, the only alternative in philosophy preceding Descartes, it is a striking fact that contemporary epistemologists ignore it so often. Ever since Aristotle it has been traditionally recognized in epistemology that the prime candidate for a domain of justified knowledge claims should be the body of scientific claims. Logical empiricists characteristically admit that the question of justification in science can only be answered by analyzing the logical structure of science (Salmon 1963a, 341-43). This view of *logical empiricists* suggests that the most appropriate kind of the theory of justification to be developed in the course of this logical analysis of science would be a semantical theory. And in fact in the rest of this book I am going to focus almost exclusively on the semantical kind of the theory of justification. The semantical character of such a theory is not its generic feature, however. K. Ajdukiewicz, for instance, provides a framework for developing the same kind of the theory of justification, which is, nonetheless, a pragmatic theory (Ajdukiewicz 1965).

This seems a totally different tradition of approaching problems than the one suggested by definition of mentalistic theory of justification. Nevertheless, the two approaches to justification could well be made continuous. R. Wójcicki, for instance, demonstrates that on the grounds of a minimalist theory of justification it is still possible to provide a mentalistic theory of scientific knowledge (Wójcicki 1995/1996a, 343-44). In my paper "Two Concepts of Justification" (Kawalec 1997a), where I identified and distinguished two kinds of theories of epistemic justification, I used the term "methodological" to refer to the kind of theory of justification projected by logical empiricists. The term "methodological", however, is not apt here for it seems to be misleading, and in this context it would perhaps be doubly misleading. For, on the one hand, epistemologists refer therewith to a set of practical rules applying the theoretical-epistemological considerations to a specific knowledge situation:

> *[M]ethodology* is like engineering or like the lore of carpentry. Such learning or lore is directed at certain special ends or goals (other than that of knowing more in general), and a body of such learning is a set of specified means/ends relationships (Sosa 1991, 245).

On the other hand, however, Carnap uses the term "methodological" having in mind all issues concerning the application of logical theories (Carnap 1950). Moreover, this usage is also acknowledged by epistemologists, since on this view: "[e]pistemology is the study of methodology, and proper methodology is the province of deductive logic, inductive logic, probability theory, and statistics. Epistemology reduces to these subjects, all of which are *formal* disciplines. *Empirical* sciences are not needed" (Goldman 1986, 2).

Therefore, instead of "methodological" I propose here "minimalist". For the theories of justification motivated by the condition (MITJ), that I state below make minimal ontological commitments concerning the existence of as well as the ontic nature of justification-bearers such as mental states. Thus we have:

DEFINITION 2.6 *Minimalist theory of justification* (MITJ)

T is a minimalist *theory of the epistemic concept of justification only if the term "justified" occurring in T is a relational term defined over a set of sentences* $\mathbf{P}_t$ *which is a subset of the set of sentences* $\mathbf{P}$ *that belong to the body of science* (*where t ranges over the entire history of science*), *and* $\mathbf{P}_t$ *is closed under the operation of logical consequence.*

Definition 2.6 is vague in several respects. It does not tell us precisely how to determine $\mathbf{P}_t$ given $\mathbf{P}$; apart from the fact that we can expect that the set of sentences $\mathbf{P}_t$ includes such kinds of sentences as hypotheses or sentences reporting the results of experiments, i.e. evidence. Neither does (MITJ) promise to give a delineation between $\mathbf{P}$ and other sets of sentences. In this respect, however, definition 2.2 does not seem to be more explicit than definition 2.6. What is then the advantage of (MITJ) over its mentalistic counterpart? At the face value, definition 2.6—in opposition to definition 2.2—does not overtly acknowledge any psychological commitments. Precisely this point constitutes the grounds of Goldman's objection against minimalist epistemological theories. His criticism, however, is conditional on the fact that "[t]he notion of justification is directed, principally, at beliefs" (Goldman 1986, 4), i.e. is conditional on embracing a mentalistic theory of justification. The latter condition, even if drawing upon idealized beliefs, need not, however, be satisfied by minimalist theories of justification. By contrast, minimalist theories turn on the distinction between how beliefs are generated or acquired, and what is their inferential structure (Audi 1998, 155).

In fact, (MITJ) seems to afford a theory more psychological flexibility than (METJ) as it refrains from specifying the range of attitudes towards the sentences contained in $\mathbf{P}_t$ from which to pick out the appropriate ones. Still, definition 2.6 has to be constructed in such a way as not to preclude the possibility of grounding a psychologically plausible theory which will also be consonant with a minimalist epistemological theory of justification.

M. Pastin, for one, is in favor of proceeding from a minimalist theory of justification—which he calls the theory of "propositional justification"—to a mentalistic theory in view of the fact that "'psychologically real' inferences correspond to justified inferential relations among propositions" (Pastin 1981, 216). What is objectionable in his approach, however, is that such a mentalistic theory would presuppose "empirically ideal believer", violating thereby (QT). Moreover, the conditions that Pastin imposes on "empirically ideal believer" are themselves debatable (Lehrer 1981b, 239-40). In general, it is widely assumed that the transition from a minimalist theory of justification to a mentalistic theory turns on an account of the relation between propositions and belief contents (Audi 1998, 153; Marciszewski 1972, 135) rather than psychological idealizations.

As a matter of fact, we cannot afford to ignore that actual epistemic agents, indeed, foster propositional attitudes towards scientific sentences. I refer here to "sentences" rather than "propositions" for two reasons. The first is that the minimalist theories that I am going to present in the ensuing refer primarily to "sentences", although these theories do not turn on referring to "sentences" rather than "propositions". The second is that referring to "propositions" would engage me in ontological and linguistic problems that are irrelevant for the purpose at hand. (MITJ), in opposition to (METJ), does not limit the account of propositional attitudes to be cast only in terms of full-beliefs; it allows for much less confident (Kaplan 1996, 102-54) or doxastically-imported psychological attitudes (Kawalec 1997a): (MITJ) permits that doxastic and non-doxastic attitudes be kept separately. Indeed, (METJ) and (MITJ) seem to fit nicely with some epistemological distinctions, for instance between different senses of "inference":

[t]here are, then, two sorts of things involved when I draw a conclusion. One is the mental process of my concluding it on the basis of one or more beliefs of mine, as where I conclude that Alberto has been bitten by a deer tick on the basis of my belief (just acquired) that he has lime disease and my background belief that this is caused by such tick bites. The other element in my drawing a conclusion is the set of two or more propositions which are my conclusion and my ground for it. Call the first item the *inferential process*; it is a mental episode of reasoning. Call the second its *inferential content*; it is abstract and not a process" (Audi 1998, 153).

In the previous section we observed that (ST) and (QT) are in conflict conditional on (METJ). Admittedly, (MITJ) as a meta-level constraint, is more promising in the construction of an epistemological theory of justification, for it supposedly alleviates the tension between (ST) and (QT). For (MITJ) provides $\mathbf{P}_t$ as a new domain, with respect to which T has to be quietistic. And it is precisely this domain that has been traditionally recognized as one where all sorts of "doxastic, metaphysical, modal, semantic, or syntactic" expressions, classified by Goldman as non-epistemic, are naturally located. If all these non-epistemic concepts are at home with $\mathbf{P}_t$, then no problem similar to Jones's evaluative standards will emerge.

The change in the domain introduced by definition 2.6 seems to lead to a problem for (ST) and (NT) taken together, however. In the case of definition 2.2 the (ST) and (NT) conjunction seemed sufficient to yield an explanatory and thus an interesting epistemological theory of justification. Now, we will face the problem of how to discern between competing theories, which have comparative explanatory status. However, since I am not going to give a clue to how this problem might be solved in general, I will come back to it in the next chapter discussing Carnap's solution of it.

Before I proceed further, it should be emphasized that in the ensuing I will refrain from dealing with what Goldman calls a "base-clause", and what is commonly recognized as *direct justification.* I restrict my further investigations to *inferential justification* (Ingarden 1962; Goldman 1986, 83-84). Admittedly, as K. Lehrer suggests, the broad characterization of an epistemological theory, as foundationalist or coherentist for instance, does not turn on this distinction (Lehrer 1974, 210).

Inferential justification may be broadly characterized as justification which is afforded by a sentence, proposition or belief. *Direct justification* is thus justification afforded by, for instance, a fact, process, state of affairs, etc., i.e. by something which, while perhaps not devoid of propositional content, is not itself a sentence, proposition or belief (Hempel 1961, 149-50); (Hetherington 1996, 140).

The distinction between direct and inferential justification is especially problematic as regards a mentalistic theory of justification,[20] for what counts as directly justified in one context or what counts as directly justified conditional on one person's background knowledge, may count as inferentially justified in a different context or on another person's background knowledge: "we cannot in general specify propositions which can be believed only in one way, or determine whether a belief is inferential by considering just the proposition believed" (Audi 1998, 157).

With the restriction to theories of inferential justification at hand I will now proceed to state the two most important conditions for a structural reliabilist theory of justification. Accordingly, these conditions articulate what does it mean for a sentence—tautological or non-tautological—to be justified.

Suppose we have two sentences, which are members of $\mathbf{P}_t$: "All animals which have a heart have a liver as well" and "The dog called 'Aldi' has a heart and a liver". If we are to elaborate the justificatory standards starting from sample sentences like these, we need to bear in mind the (NT) condition, which prescribes us not to stick to closely to the initial sample. For there are many possible content-driven descriptions of the two sample sentences, but which, still, will fail to constitute a generic set of conditions of justification. For instance, we could modify (ST) thus:

(ST') Theory *T* of the concept of justification *J* is *substantive* iff the necessary and sufficient conditions *T* gives for *J* are stated in anatomic terms.

(ST'), however, would even be less satisfactory than (ST), since "anatomic" is only a very narrow and appears as an arbitrary subclass of "non-epistemic". We thus realize that the conditions stated by (ST') are neither necessary nor sufficient for a theory of justification. They are not necessary for the reason just stated, i.e. the anatomic terms form only a subclass of "non-epistemic" terms, and we have no good grounds to repudiate the charge that it is an arbitrary subclass. However, it is only a manifestation of a more general problem for any content-driven specification of a subclass of "non-epistemic" terms. For, on the one hand, there might be several competing subclasses, like "anatomic", "physiological", "physical" or even "functional", each of which might be claimed to represent a plausible alternative, but each of which perhaps would be confined to a very limited and often arbitrary range of cases. On the other hand, however, there might be good candidates to subsume "non-epistemic", which would nonetheless yield a different level of generality, or world-slicing, e.g. micro- and megascopic level of description. This problem, however, does not seem to be exclusively confined to (ST'), but would rather concern any similar content-driven principle.

The offer of providing a content-driven specification of sufficient conditions within (ST) fares no better. A prima facie objection is that even if this route can be successfully followed, it will yield a version of (ST) incompatible with (NT). For it is clear that sufficient condition expressed in "anatomic" terms will not be a sufficient condition for all kinds of sentences, which are members of $\mathbf{P}_t$, e.g. sentences stemming from physics or history. Having figured out what kinds of terms there are to cover all sorts of sentences, which are the members of $\mathbf{P}_t$, we will end up with a great number of sufficient conditions for the kinds of terms in question. This, in turn, will impose a great number of partition-cells upon $\mathbf{P}_t$, thus violating (NT).

Another reason why content-driven specification of sufficient conditions for (ST) will not grant the theory *T* in question the "substantive" character is this: suppose that we can carry out the content-relative partition of $\mathbf{P}_t$, and by the same token the partition of the sufficient conditions of (ST), in a way which accords well with (NT). Each of the partitioned sufficient conditions will amount to acknowledging that *T* is "substantive" for this part of the discourse, which is discriminated by the appropriate kind of terms within $\mathbf{P}_t$. Suppose further that each sufficient condition is a legitimate one, e.g. that the sufficient condition stated in "anatomic" terms is a sufficient condition for *T* to be substantive. Since the sufficient condition cast in "anatomic" terms does not have a universal applicability for all kinds of discourse, it implies that *T* is substantive only with respect to the kind of discourse which is discriminated by "anatomic" terms. This sufficient condition tells us nothing, however, about *T*'s being substantive in other kinds of discourse beyond the "anatomic" one. What this seems to endorse then is a discourse-relative concept of "substantive" theory; such a concept, as we have already noticed, would undermine the nomic character of the theory in question.

An alternative to the foregoing ways to modify (ST) might be a content-independent formulation of necessary and sufficient conditions for justification continuous with (MITJ). This could, for instance, be a modification which would turn on the structure of given sentences. Thus, the initial "anatomic" sentences could read as follows: "$\forall x[H(x) \equiv L(x)]$", and "$H(a) \wedge L(a)$", or "$\exists x[H(x) \wedge L(x)]$"; where "H" stands for "has a heart", "L"—"has a liver", and "a"—"Aldi". The seeming advantage of this approach over the content-driven one is that it is universal across different content-specific kinds of discourse, unless we do not have appropriate logical means to cope with the structure of sentences. While consenting to the irrelevance of the content of sentences, we can afford a uniform set of "structural" terms to provide the necessary condition within (ST):

DEFINITION 2.7 *Structurally substantive theory* (STS)

A minimalist theory of justification T is substantive *only if T states necessary conditions for justification in structural terms.*

It is rather straightforward that (STS) is continuous with (ST), since we appreciate that the building-blocks of "structural terms", i.e. "modal, semantic, or syntactic expressions are not epistemic". Still, there may more than one way to elucidate the meaning of the term "structural" occurring in (STS). Provided we limit the discussion as pertains to inductive logic, there still seem to be two major alternative approaches—as was in some detail illustrated in chapter 1. One is to provide a mixed representation of probability space, combining the standard probability theory with semantical concepts (esp. the concept of model as a point in probability space, and logical vs. factual truth; cf. Carnap 1971a, 35-67). The other approach is more confined to the categories of the standard probability theory (see e.g. section 4 of the first chapter; Kuipers 1978b; Howson and Urbach 1993; Zabell 1997). Of course, this is not a trivial difference, and I leave it for the moment as an open question whether the difference between the two approaches is also principal, but still, the former is apparently more apt to structurally model logical relations between sentences (propositions). And if we insist on sustaining the distinction between logical and empirical probability, then the latter approach could be reserved to model instances of carrying out factual inductive inferences.

As regards the former approach, it might appear that "structural terms" would be severely limited because of the confined vocabulary of the languages studied in the systems of inductive logic. This worry, however, may be somewhat alleviated by outlining the results which could be seen as continuous with the discussed approach. For instance, H. Gaifman and M. Snir (Gaifman and Snir 1982) set out a fairly complex structure for a language, which is rich enough to prove J. L. Doob's basic martingale convergence theorem (roughly, if the evidence is sufficiently complete, then almost surely the posterior probability of a hypothesis h approaches 1, if h is true).[21] A highly sophisticated project continuous with the systems of inductive logic presented in chapter 1, is brought out by R. Chuaqui. In a similar vein, he attempts to explicate the concept of probability as "the degree of possibility of truth", drawing upon both semantical concepts (e.g. model and truth) and a modified principle of indifference (cf. Chuaqui 1991; 1994). A general representation theorem is proved to the effect that any stochastic process can be approximated, up to an infinitesimal, by this "relational" variety of structural representation espoused by Chuaqui.

This outline is meant to rule out some important alternatives[22] rather than to provide a precise indication on how to pick out (systems of) structural terms. And perhaps none of the authors discussed in this book will claim that at the present stage of the development of science such a resolution could be accomplished. Yet, in most philosophical applications the choice of either of the above delineated frameworks of structural terms—logically amplified or standard probability theory—does not seem to effect in thoroughly incompatible representations. Therefore, in the definition to follow I will broadly refer to "a language system", meaning thereby not specifically Carnap's own constructions, but rather one of the frameworks introduced in chapter 1 and outlined in this section. Some of them, of course, would require according modifications of the definition.

And still, the important feature these frameworks have in common is they do not contain epistemic terms. Of course, it is remains an open possibility that epistemic terms supervene on structural terms—and this feature would be highly appreciated were anyone to proceed from the structural theory of justification to a full-blooded theory of knowledge. Though, the appropriate choice of a framework of structural terms allows not to conflate the two kinds of theories.

Regardless of the choice of the framework of structural terms, the prevalent problem is how to proceed with the construction of the sample space. Allowing for some flexibility in this regard, sentences of the set $\mathbf{P}_t$ may in general be claimed to attain structural translation as follows:

DEFINITION 2.8 *Structural translation*

Let **M** *be the set of all models of a given language system L* (*with* **ST** *as the vocabulary of L*) *and* $\mathcal{F}$ *a σ-field generated by the set of atomic sentences of L on* **M**.

A sentence $^{\mathrm{ST}}p$ *of L is* a structural translation *of a sentence p of* $\mathbf{P}_t$ *iff there is a set of models* M ∈ $\mathcal{F}$ *in which* $^{\mathrm{ST}}p$ *is true and* $^{\mathrm{ST}}p$ *is obtained from p after every term occurring in p which does not belong to* **ST** *nor can be explicitly defined by terms which belong to* **ST** *is replaced by a term defined by terms which belong to* **ST**.

A number of remarks complement this definition. Of course, the sentences which express a measure defined on $\mathcal{F}$ are also regarded as structural, though they are not included in structural translation. This is due to the fact that—on the structural theory of justification—non-structural judgments of the degree of justification supervene on the probability measure (confirmation) of the relevant sentence, and not the other way round. So, while a mentalistic theory of justification continuous with its minimalist counterpart will perhaps come up with a rule specifying how to transform the degree of confirmation of a given sentence into appropriate epistemic conditions, there seem to be no need for a reverse rule. A natural context to broaden the above definition of structural translation in order to accommodate this sort of reverse rule would be constituted by meta-level theories of justification. The latter, however, surpass the limits of this book.

Another point concerns the concept of model employed in the definition of structural translation:

> For abstract mathematical theories, a slightly different form of models is customary, one in which the specification of a model includes a specification of a domain of individuals. In such theories, not only the kinds of individuals may vary from model to model, but even the cardinal number of the domain. ... This seems useful for abstract (uninterpreted) axiom systems ... [I]n inductive logic it is more useful to specify for each investigation, once for all, the domain of individuals and ... the partition of the value space to be used, so that these specifications hold for all models, that is for all possibilities considered (Carnap 1971a, 56).

This should not, however, subsume or conflate the difference between model and interpretation.

The other remark is the following. It is possible to imagine that at some stage of the development of science the sentences of $\mathbf{P}_t$ closely resemble their structural counterparts. Notwithstanding this similarity, the question whether ${}^{ST}p$ is true (in a given model) "is done by a purely logical analysis" of this sentence and the relevant model "without any use of factual information" (Carnap 1971a, 57). This is pertinent to the structural theory of justification espoused in the systems of inductive logic, because formal considerations alone suffice to determine which models a given sentence is true in, and the task of determining degree of confirmation of a given sentence may be straightforwardly relegated to a measure function defined on models.

The issue left open by the definition of structural translation is whether there is only one language to serve as a basis for structural translation of all sentences which belong to $\mathbf{P}_t$? If the question concerns the present state of affairs, the answer should perhaps be in the negative. Would it, however, be the same, given the whole history of the universe? An answer to the latter question—as attainable on the grounds of the systems of inductive logic—as well as the related issue whether parameters, like α, γ, κ, or λ are structural terms—is studied in some more detail in section 4 of chapter 3.

Regardless of the questions mentioned in the preceding paragraph, there is still a question whether structural translation is unique? The history of the systems of inductive logic, as evidenced to some extent by chapter 1 of this book, provides a good ground for scepticism. Even though, we would still require that the degree of confirmation be invariant as regards different structural translations of sentences which belong to $\mathbf{P}_t$ (recall esp. propositions 1.8 to 1.12, and definition 1.11). One way to pose this requirement in more precise terms was already discussed in the form of the criterion of linguistic invariance (see page 29).

Admittedly, it is beyond the present concern to be more specific about how precisely the process of structural translation is carried out. This specification can be provided in various ways, and in the previous chapter I discussed two different kinds of procedures carried out by Carnap and Hintikka. For that purpose Carnap invents the concept of state-description, while Hintikka develops the concept of attributive constituent.

Now we proceed to the question whether it is possible to render a set of sufficient conditions of the "substantive" character of an epistemological theory of justification cast in "structural" terms. Such a route, it seems, might be possible conditional, however, on the account of the success of the application of structural terms to scientific investigations.[23] Presumably, it will assume the following form:

DEFINITION 2.9 *Substantive minimalist theory* (SMT)

A minimalist theory of justification T is substantive *iff T states necessary and sufficient conditions for justification in structural terms.*

With (SMT) at hand we can proceed to formulate a necessary condition of justification, which will also establish an essential part of a structural reliabilist theory of justification. As previously stated no counterpart of Goldman's base condition (BC) is of interest to us here. The fact that I do not provide a base condition within structural reliabilism does not mean, however, that none can be given. The reason to represent the necessary condition for justification on the grounds of structural reliabilism as a counterpart of Goldman's recursive condition (RC) is twofold. On the one hand, structural reliabilism is concerned with a concept of justification which denotes relational properties of sentences. On the other hand, in opposition to Goldman, I do not presume that inferential justification is merely a transmission of the justificatory status from one justified sentence to another (Audi 1998, 159).

The necessary condition for justification is supposed to satisfy the condition for the minimalist theory of justification (MITJ), and has to be substantive and quietistic in the sense specified by (SMT) and (QT). Hence, we obtain the following recursive condition:

DEFINITION 2.10 *Recursive structural condition* (RSR)

For any two sentences p_1 and p_2 which are the members of the set of sentences $\mathbf{P}_t$, p_2 is justified *by p_1 only if p_1 is structurally translated as ${}^{ST}p_1$, and p_2 is structurally translated as ${}^{ST}p_2$, and ${}^{ST}p_2$ is justified by ${}^{ST}p_1$.*

Of course, the locution "${}^{ST}p_2$ is justified by ${}^{ST}p_1$" can be defined in different ways. Therefore (RSR) does not determine a unique structural reliabilist theory of justification. Nevertheless, I limit my discussion to follow to the problem of how the Carnap-Hintikka style systems of inductive logic define this locution. Admittedly, the viability of the structural reliabilist recursive condition for justification (RSR) turns on this problem. In the next two sections to follow I will present Carnap's and Hintikka's epistemological motivation underlying their defense of (RSR).

4. Carnap's Structural Reliabilism

The main task of this and the next section is to apply the framework of the structural reliabilist theory of justification to the Carnap-Hintikka style systems of inductive logic. The development of the systems of inductive logic which I discussed in chapter 1 suggests that the most natural way to proceed with this task is to commence with an analysis of the theory of justification represented in the systems proposed by Carnap (this section), and then to undertake a similar task with respect to the construal advanced by the Finnish School (the next section).

The theory of justification represented in Carnap's systems of inductive logic and the Finnish School's systems thereof exhibit essentially the same traits, although Carnap and the Finnish School disagree as regards the kind of theory of knowledge which is to be based on this theory of justification. It is a convenient practice in contemporary epistemology to keep theories of justification and of knowledge separate, and I follow this procedure throughout this book. Moreover, as the title indicates, I do not attempt to examine the characteristics of the respective theories of knowledge, but to limit my study to the theory of justification.[24]

The Bayesian reading of the Carnap-Hintikka style systems of inductive logic (see section 4 in chapter 1 and section 1 in chapter 3) purportedly gives reason to claim that Carnap and Hintikka advance their own peculiar Bayesianism. Nevertheless, an attempt to pursue this argument along epistemological lines would appear rather odd from a Bayesian perspective. It seems not only the case that the general epistemological issues are not much debated in the Bayesian literature, but also more traditionally oriented epistemologists refuse to admit the relevance of Bayesianism to theories of epistemic justification.[25]

The moral for the present argument is twofold. First, the objection is that on Carnap's construal in *Logical Foundations of Probability* (Carnap 1950) and in *The Continuum of Inductive Methods* (Carnap 1952b) the epistemological problems are not overtly discussed;[26] the system of inductive logic is primarily a formal construction, and the epistemological issues are but a follow-up interpretation no more relevant to the very construction than in the case of deductive logic. Second, it is crucial that Carnap's system is semantical, which undermines the very possibility of its being a theory of epistemic justification as it does not an afford epistemic subject—the possible bearer of propositional attitudes.

Let me now discuss these objections in turn. In general, the objection that Bayesianism is not an epistemological theory, and in particular an epistemological theory of justification, seems to be a manifest misunderstanding. Take as an instance the probabilistic epistemology motivated by Bayesianism.[27] This assumes that justification guides changes in probability ascription to given propositions rather than a mere accumulation of asserted propositions. With respect to partial beliefs, as will become clearer in the ensuing, Carnap seems to share this epistemological standpoint.

In particular, the objection is also dubious with respect to the Bayesianism advanced by Carnap,[28] who does not hesitate to impose epistemologically motivated constraints on the systems of inductive logic, being fully aware of the trade-off between the epistemic merits of the construction and its formal neatness. The proponents of the full mathematization of Carnap's inductive logic apparently fail to appreciate this fact.

To give the reader some idea of the flavor of the epistemology in *Logical Foundations of Probability*[29] let me pick out some of its disclosures. Carnap persistently uses explicit definitions of relevant concepts rather than axioms, which would have constituted the implicit definitions of the primitive terms. For a logicist like Carnap it is only the first procedure that warrants *the logical construction of a given object*, while the second merely states lawful connections between objects of *any domain* that satisfy the axioms.[30] It is only the former approach to formal constructions that warrants the thesis that theorems are logical truths about well-defined objects. Carnap does not hesitate to admit that:

> While deductive logic may be regarded as the theory based upon the concept of logical consequence or deducibility, inductive logic is the theory based upon what

> might be called the degree of inducibility, that is, the degree of confirmation (Carnap 1950, 2).

An important suggestion as to how we are to appreciate this assertion are set forth by Richardson (Richardson 1998, 176), who identifies and scrutinizes a similar procedure in earlier works of Carnap. A concept can be introduced in a given system only if the precise rules governing its usage are stated. So, when in *Logical Foundations of Probability* Carnap admits that the concept of the degree of confirmation is fundamental, he seems to imply thereby that the pertinent theorems constitute the foundation of any inductive inference.

Carnap makes no effort to give a rationale for the introduction of numbers within his allegedly logical theory.[31] He is seemingly careless in giving credit to the very idea of the *quantitative* nature of inductive logic (see section 1 in chapter 1). As will be more fully appreciated below, numbers within the system of inductive logic play the fundamental epistemological role of casting the empirical data in an objective form and thus subsuming them into lawful connections specified within the system. Hence the quantitative character of inductive logic is not merely a matter of the precision of its representation, but also a matter of setting its foundation.

There are many extra logical assumptions within Carnap's system of inductive logic that permit the specification of an efficient method for the calculation of the degree of confirmation. For instance, it is assumed that there are j primitive predicates, with j being a natural number, in the language system L, in which both the hypothesis and the evidence statements are formulated. Next, it is assumed that there are n individuals in the universe, with n being a natural number. Mutually exclusive and together exhaustive conjunctions of primitive predicates, some of which are negated, form $k = 2^j$ so called Q-predicates. All possible distributions of Q-predicates among the n individuals are called *state-descriptions* (definition 1.2), which are then grouped according to how many instances of the given Q-predicate they specify, to form so called *structure-descriptions* (definition 1.12), each of which is an alternative of the state-descriptions having the same number of instances for given Q-predicates. There are k^n state-descriptions, and $(n + k - 1)! / n!(k - 1)!$ structure-descriptions. The state-descriptions are mutually exclusive and collectively exhaustive; for any formula in L its normal form is an alternative of some of the state-descriptions. The degree of confirmation is defined in terms of a one-argument measure function $m(.)$, which assigns a uniform prior distribution to all structure-descriptions, and then, conditional on the priors, a uniform distribution to all the state-descriptions constituting given structure-description. For instance, if there is one primitive predicate "lands heads" $H(.)$ and two coin tosses (a_1, a_2), then there are two Q-predicates

$$Q_1: H(.)$$

and

$$Q_2: \neg H(.),$$

four state-descriptions

$$S_1: Q_1(a_1) \wedge Q_1(a_2);$$

$$S_2: Q_1(a_1) \wedge Q_2(a_2);$$

$$S_3: Q_2(a_1) \wedge Q_1(a_2);$$

$$S_4: Q_2(a_1) \wedge Q_2(a_2);$$

and three structure-descriptions:

$$Str_1 = S_1;$$

$$Str_2 = S_2 \vee S_3;$$

$$Str_3 = S_4;$$

where

$$m(S_1) = 1/3;$$

$$m(S_2) = m(S_3) = m(Str_2) / 2 = 1/6;$$

$$m(S_4) = 1/3.$$

Although it is often found objectionable that Carnap limits in practice his systems to finite k, the following principled motivation is hardly mentioned:

> if... we think of the world as conceived in theoretical physics, then we find a finite and, indeed, a very small number of fundamental magnitudes to which, according to the assumption of physicists, the great variety of phenomena is ultimately reducible (Carnap 1950, 75).

Note that *the finite variety assumption* (see also page 10) endorsed here by Carnap is further exposed by the role it plays within his system of inductive logic. The finite variety assumption applies in the first place to primitive predicates (or properties) yielding thus an efficient method to generate all basic statements in *L*, i.e. the state-descriptions, to which all other statements expressible in *L* reduce. If we further assume, quite correctly as will be argued later, that the statements thus generated bear on the epistemic status of knowledge states of epistemic agents, then a version of epistemic foundationalism seems inevitable.

Carnap's alleged foundationalism seems to be common knowledge (Campbell and Paller 1989, 235). It apparently is a common view that the program initiated by Carnap's *Logical Foundations of Probability* was motivated by a desire to restore the old foundationalist paradigm of "perfect, infallible knowledge" in the face of the fact that "it was clear that theories could not be classified as provably true or false"; thus, "Carnap thought that while science was conjectural, the theory of probabilistic confirmation would be *a priori* and infallible..." (Lakatos 1968, 317-18).[32] Most notably, Hempel asserts that Carnap's interests were

> united by a common *Leitmotiv*, namely, the search for ... reformulations, or explications, of the basic idea of empiricism that all our knowledge of the world ultimately derives from what is immediately given to us in the data of our direct experience (Hempel 1973, 57).

P. Suppes comments on this foundationalist picture of *Logical Foundations of Probability* as pursuing "the misguided tradition of logical atomism" of Hume's *Treatise* and Wittgenstein's *Tractatus* to its more complete and constrained probabilistic version (Suppes 1966a, 202). In his criticism Suppes draws upon the fact that the systems of inductive logic force one to express the entire background knowledge, which is highly unrealistic. Such a procedure, nevertheless, is epistemically advantageous (Audi 1998, 154).

In the same foundationalist vein[33] the construction of a language projected by Carnap implies that we can codify all human knowledge in a given formal structure, but also similarly codify all possible relations between non-tautological sentences. The latter as a necessary condition requires that the set of basic statements built out of the primitive predicates and individual constants, and then combined to exhaust all possible combinations to form the set of state-descriptions, represents all extra-linguistic facts or states of affairs. This assumption is incorporated within the system of inductive logic rather smoothly via the finite variety assumption and on the grounds of Carnap's general epistemology in *Logical Foundations of Probability*.

The preceding discussion seems to bring out a foundationalist picture of the epistemology in *Logical Foundations of Probability*, satisfying the foundationalist stipulations:[34] there is a set of mutually exclusive and collectively exhaustive statements (state-descriptions) with the established epistemic status, upon which all other statements depend. This seemingly foundationalist leaning of Carnap's epistemology in *Logical Foundations of Probability* is essential to the entire formal construction, since without it the latter loses its rationale as a foundation for inductive inferences.

K. Lehrer, for one, advocates a foundationalist reading of Carnap's inductive logic (Lehrer 1974, 138) which is supposed to be grounded, however, in – what is recognized to be mistaken – Carnap's purported empiricism (Lehrer 1974, 77). R. Chisholm, on the other hand, claims that epistemological foundationalism does not turn on the issue of logic, either deductive or inductive, for it deals with a different kind of questions:

> in addition to the "rules of deduction" and the "rules of induction," there are also certain basic "rules of evidence." The deductive logician tries to formulate the first type of rule; the inductive logician tries to formulate the second; and the epistemologist tries to formulate the third (Chisholm 1966/1977, 3).[35]

If, however, one tends to adopt the foundationalist reading of the general epistemological position of Carnap in *Logical Foundations of Probability*, then what is at odds with this view is Carnap's continual insistence on those alleged foundations of all empirical knowledge having a purely formal and analytical nature. Characteristically the similar point is advanced by W. Sellars, who develops a minimalist theory of justification prior to Carnap: "I share the conviction that all justification in the mentalistic sense of the term is ultimately analytic in character" (Sellars 1947, 648 footnote 4). The crucial step here is then to interpret the primitive predicates in terms of observational predicates. The fact that Carnap does not provide a specification of the basic statements in terms of content—despite the semantical character of the construction—but limits himself to structural specification accords well with the interpretation of his general epistemological position in *Logical Foundations of Probability* I advance throughout this book.

Another argument may be advanced, as suggested by Lehrer (1974, 135), to demonstrate that substantive foundationalism is incompatible with the view that deductive inference is the only method of the expansion of knowledge states. In what follows I argue that Carnap in *Logical Foundations of Probability* and in *The Continuum of Inductive Methods* is committed to the latter view.

The inadequacy of foundationalist interpretation of Carnap's epistemology motivates the exploration of an alternative epistemological position which could also account for the foundationalist leaning of Carnap's position. Below I elaborate this alternative in terms of structural reliabilism.

Although in the present section I limit myself mostly to discussion of the prior distribution, in particular the $m(.)$ measure function of *Logical Foundations of Probability*, there is an intriguing fact about the confirmation function $c(.|.)$ which I would like to emphasize at this point. Carnap's definition of the concept of the degree of confirmation, which gives rise to the definition of $c(.|.)$, incorporates the axioms of probability theory (Axiom 1)-(Axiom 3). The latter, however, is shown by Ramsey, and by other Bayesians, to be equivalent to coherence conditions.[36] If my beliefs fail to satisfy the axioms (Axiom 1)-(Axiom 3) and are thus incoherent, then, as Ramsey proved, it is possible to set up a series of bets concerning an outcome of an experiment in which, independently of the outcome, I will always loose. Within his system Carnap accommodates an even stronger version of coherence, i.e. strict coherence. My beliefs are strictly coherent if in addition to be coherent they are such as to never allow me to engage in a bet where there is no possible gain.[37]

For mentalistic coherentists a belief is justified if it coheres with the body of background knowledge; coherence is then tantamount to justification.[38] Suppose, for instance, that in my visual field I identify the same sort of experience that appears whenever it is raining. The statement that it is raining now coheres with my body of knowledge, which includes the information that it is summer, and that within the last few hours the pressure has decreased significantly.

Since Carnap puts still further conditions on the concept of confirmation beyond that of strict coherence, he might be said to appropriate coherence as a necessary, but not sufficient, condition for justification. In fact, Lehrer advances an argument, which suggests that Carnap, who embraces strict coherence, is inevitably committed to coherentism:

> If basic beliefs are not justified in a way that guarantees their truth, then in accepting the propositions in question we take some risk of error. If we take such a risk, then it must be possible to explain what makes that risk worthwhile. If we explain this, the explanation will provide a justification for accepting the basic proposition. The assumption that the belief is basic, self-justified, is either refuted by the justification, or, at best rendered otiose (Lehrer 1981b, 233).

What seems to set Carnap apart from mentalistic coherentism, however, is the fact that a necessary condition for coherence evaluation is a content specification of a given statement or propositional attitude.[39] For instance, suppose that my background knowledge is as in the above example and I am appeared to as if it is snowing. Granted this, I can predict that the latter statement is incoherent with my body of beliefs only if I have access to the content specification of the relevant statements. Here again, as in the case of his purported foundationalism, Carnap purposely fails to give any interpretation to the basic statements, and the specification is purely formal. A natural question is, then: how is it possible, on such a position, to evaluate coherence, if all that is available is the formal specification of their structure and mutual relationships within a given language system *L*? As will become clear below, the concept of coherence applies to the epistemology in *Logical Foundations of Probability* to the extent that it satisfies the conditions set up by structural reliabilism. Therefore the examination of the concept of coherence would not elucidate Carnap's epistemology in *Logical Foundations of Probability*,[40] but still it might be worthwhile to trace precisely to what extent the coherentist position turns on the content specification of the relevant statements or propositional attitudes.

In discussion of the foundationalist and coherentist traits in Carnap's construction I presupposed that it makes sense to classify his position among the alternative theories of epistemic justification. Now I shall face the objection that appears to undermine this procedure. The point which most contemporary epistemologists would be willing to accept is that the bearer of justification is belief (Alston 1989, 83). However, Carnap's system of inductive logic is a semantical construction. This precludes such concepts as propositional attitudes. Therefore the prima facie epistemological worry is that Carnap's theory of the degree of confirmation in *Logical Foundations of Probability* and *The Continuum of Inductive Methods* does not constitute a genuine epistemological theory.

Goldman's articulates this objection as follows:

> degree of belief should be identical to weight of evidence. It is normally assumed that weights of evidence are derivable from certain formal facts together with the cognizer's present evidential corpus. The critical formal facts are either of a semantic nature—so Carnap viewed statements of inductive probability, or confirmation theory... Logical implication is a special case of such formal facts, where the premise proposition confers a confirmation of degree 1 on the conclusion.
>
> If this approach is correct, then all talk of "processes", especially psychological processes, seems to be irrelevant... [F]ields like statistics and confirmation theory cannot *by themselves* give us an adequate theory of justifiedness (Goldman 1986, 89-90).

A similar objection is put forward by A. W. Burks in (Burks 1963, 739). This kind of objection, however, is based on a confusion of mentalistic and minimalist theories of justification, which Carnap sharply distinguished (see e.g. the methodological rule (MR), definition 2.11 below).

I am inclined to accept Goldman's point to the extent that if we adopt the standard view that knowledge entails belief, then it is hard to imagine how the type of theory of justification Carnap projects in *Logical Foundations of Probability* and *The Continuum of Inductive Methods* could directly yield a definition of this concept of knowledge (Lehrer 1974, 12).[41]

I will begin my defense of the epistemological character of the systems of inductive logic with the following observation. In the epistemological account of why an epistemic agent *S* is justified in believing in *p*, where *p* is logically entailed by her other beliefs *q* and *r*, it is a necessary condition that the actual inference made by *S* conform to a schema of valid deductive inferences. The point is that deductive logic in contemporary epistemology plays the role of determining the epistemic canons against which to assess the status of some of the epistemic states of an epistemic agent (Audi 1998, 163-64). Deductive logic, however, does not set up epistemic canons for non-demonstrative inferences; in *Logical Foundations of Probability* it is inductive logic that takes on this task, i.e. Carnap's system of inductive logic is intended to provide foundations for all traditionally recognized forms of inductive inference. In this view deductive and inductive logic together form an exhaustive set-up of epistemic canons for all kinds of inference. These canons do not form a set of sufficient conditions for justification, because empirical epistemic agents may, for various reasons, fail to make actual inferences between beliefs they hold even where epistemically licensed inferential connections obtain.[42] This set of epistemic canons plays a crucial role among the necessary conditions for inferential justification, for other conditions will only concern whether a given set of epistemic canons specified by logic applies to a given knowledge state of an epistemic agent.[43]

The Goodman paradox seems to countenance precisely this point for "we cannot base justificatory inference on the form of the inference alone" (Lehrer 1974, 133). A similar argument is advanced by Sosa (Sosa 1991, 127; Lycan 1998). In the next chapter I argue that this kind of argument fails as regards Carnap's construal, because it does not appreciate the meaning that the locution "the form of inference" acquires in Carnap's account.

From Carnap's perspective, however, there is an important difference between how the epistemic canons set up by deductive and inductive logic operate. The former results in the "acquisition of a new sentence", while the latter leads to "a determination of its degree of confirmation" (Carnap 1950, 202). This difference—due to the fact that within inductive logic there are no additional "inductive" rules of inference besides those of deductive logic[44]—introduces a heterogeneity into the set of epistemic canons.[45] The ones set up by deductive logic control the expansion of knowledge states in respect of the transmission justificatory status.[46] The canons set up by inductive logic extend beyond the former concerning the evaluation of the epistemic status of meta-level compound sentences of the form "$c(h \mid e) = r$", where *h* and *e* are sentential input arguments.

The heterogeneity within the set of epistemic canons elucidates Carnap's epistemology in *Logical Foundations of Probability*. Since the former was supposed to embrace norms for all epistemically granted inferences, it follows that the only kind of inference whose justificatory status is epistemically licensed in Carnap's account is deductive inference. The word "inference" as attributed to "inductive" merely accounts for the difference between different values a hypothesis is assigned on the changing evidence. Properly speaking it is a misuse of the word "inference" in this context, as Carnap concludes (Carnap 1950, 206), because it does not, in effect, yield any new knowledge states. This is why in *Logical Foundations of Probability* inductive inference operates in clearly internalist fashion, which is in the meta-level evaluation of epistemic states.[47] In this view Carnap's refusal to accommodate acceptance rules within the systems of inductive logic gets its epistemological credit. Carnap was perfectly aware that propounding any sort of acceptance rule would lead him to embrace foundationalism (Hempel 1963, 702-703).[48]

The internalist motive raised in the picture of the epistemology of *Logical Foundations of Probability* is instructive for interpretation of the following methodological rule:

DEFINITION 2.11 *The methodological rule* (MR)

Suppose that inductive logic supplies a simple result of the form "$c(h\,|\,e) = r$"... If e expresses the total knowledge of X at the time t, that is to say, his total knowledge of the results of observations, then X is justified at this time to believe h to the degree r, and hence to bet on h with a betting quotient not higher than r (Carnap 1950, 211).

This methodological rule does not advance the knowledge claim with respect to *h*, for it is only "$c(h\,|\,e) = r$" that is epistemically licensed by inductive inference. Similarly, inductive inference does not enable an epistemic agent to enhance the body of evidence *e*. Therefore *h* and *e* are relevant to Carnap's theory of justification in as much as they conform to the construction rules of a given language, enabling thus the evaluation of $c(.\,|\,.)$. Note, however, that Carnap reveals here also his bias towards probabilistic epistemology in attributing the justification, but not knowledge, to *h*.[49]

However, the above quotation must be perplexing for an advocate of the empiricist foundationalist stance on Carnap's epistemology in *Logical Foundations of Probability*. For it is apparent that Carnap takes the formula "$c(h\,|\,e) = r$" to be sufficient for determining a fair betting quotient for *X* in his actual betting behavior. Similarly, "$c(h\,|\,e) = r$" is sufficient as the frequency estimator. However, both these claims appear somewhat paradoxical when we consider that the system of inductive logic, which generates the assignment of *"r"* to *"c*(*h* | *e*)", is purely formal and analytical, based on the uniform prior distribution among the structure-descriptions of *L*, in which *h* and *e* are formulated. On the other hand, however, in Carnap's account the mathematical function (see definition 1.17 on page 24) that characterizes the predictive probability (of the next event to instantiate *i*-th *Q*-predicate) forces us to treat logical formulae as if they were events susceptible to empirical observation. In order to determine the value of the function G Carnap designs specific rules which guide the "observation" and counting of the number of occurrences of particular predicates in a logical formula (Carnap 1952b, 11-18).

A closer look at Carnap's construction in *Logical Foundations of Probability* will, I suggest, clarify the view on what kind of epistemological theory he advocates. A comprehensive picture of Carnap's epistemology in *Logical Foundations of Probability* would require a coherent account of how the basic statements are generated, and what is precisely their epistemic status. In (Kawalec 2000b) I make recourse to Carnap's ontology in *Logical Foundations of Probability*, which seems to account well for how the basic statements make their appearance in the construction. Nonetheless, it might also be worth to study Carnap's purported reliabilist position on that matter, yielding thus a thorough reliabilist account. This issue, however, is beyond the present concern, as it does not affect Carnap's position on non-inferential justification.

As was already made clear, Carnap's only epistemic canons of inference are specified by deductive and inductive logic. Let me now focus on how the structure of inference is specified by the latter. We have already observed some extra logical assumptions in the construction of *L*. There are also some metric conventions governing the behavior of *m*(.); for instance, it takes on the value 0 for the conjunction of all state-descriptions in *L* (counter-tautology), 1 for the disjunction of all state-descriptions in *L* (tautology), and assigns equal values to structure-descriptions rather than to state-descriptions as was the usual practice in the classical probability theory (see chapter 1, especially page 10). The metric conventions play a twofold role in *Logical Foundations of Probability*.[50] First, they enable one to define within the system of inductive logic an efficient method for the calculation of the degree of confirmation for any given sentences of *L*. Second, they tie up inductive logic with empirical data cast in quantitative terms.

Carnap has several arguments in *Logical Foundations of Probability* that support foundational research on inductive inference in terms of logic and quantitative structure:

> if we want to obtain knowledge of things and events of our environment..., then the roundabout way which leads first away from these things to an abstract schema may in the long run be better than the direct way which stays close to the things and their observable properties (Carnap 1950, 217).

Here Carnap articulates the view that his primary epistemological motivation relies on the past record of the success of science. In the view of this success it is appropriate to set up the set of scientific sentences $\mathbf{P}_t$ as the proper domain of the epistemic concept of justification, for instance, in the way suggested in (MITJ) (see definition 2.6 on page 67).

Carnap characteristically refers to the example of geometry as a model of how a substantive theory should proceed. The purely structural, as opposed to qualitative, specification and elaboration of geometry is epistemologically advantageous:

> Although the aim remains the investigation not of the abstract configurations but of the observable spatial properties of concrete things, nevertheless it turns out that abstract geometry supplies the most efficient method for this investigation, much more efficient than any method dealing directly with observable spatial properties (Carnap 1950, 216).

The crucial step in the development of geometry was to detach it from "the world of concrete things with their observable properties" and make it "a schema consisting of constructs..., transformed into a theory of certain spatial configurations whose properties are completely and exactly determined" (Carnap 1950, 216). This step enabled later developments: transformation of geometry into the theory of sets of real numbers, into a formal axiom system, and finally into a "special branch of the logic of relations". The latter steps permitted the development of physics.

The important epistemological moral for Carnap is that inductive logic, too, should be purely structurally specified, and determined as a semantical-formal discipline.[51] Thus, what from Carnap's perspective appears to warrant the success of geometry, constitutes via the foundations of inductive inference in *Logical Foundations of Probability* the epistemic canon for any inductive inference whatsoever. The core conviction is that the sufficient condition for determining the epistemic status of a statement is a structural specification of the language system in which the statement is expressed, coupled with the mutual relations of all other statements of the language. Thus the moral drawn by Carnap from the analysis of what constitutes the substantive character of geometry is that it is a logical structure embedding conventions strong enough to include numbers within the construction, which is sufficient to warrant the transmission of epistemic merit via inferences. According to Carnap's perspective on science as "unified", if epistemological theories, in particular epistemological theories of justification, are to participate in the success of science, then they have to be substantive theories in the same sense as geometry. This, I think, is Carnap's basic argument to the effect that an epistemological theory of justification should be substantive in the structural sense elucidated in (SMT).

There is still another argument which Carnap develops to defend the view that structure is sufficient for determining epistemic status of scientific sentences. It might be objected, Carnap admits, that structural specification leaves "genuine qualities" or "qualia" "forever outside our grasp". Carnap rejects this claim, and illustrates his standpoint by means of the following analogy. There is no finite procedure to cover a circular area with non-overlapping quadrangles drawn within the circle. Still, it is not true that there is a point that cannot be covered except by using a different method.[52] The moral for logic is the following:

> In any construction of a system of logic or, in other words, of a language system with exact rules, something is sacrificed, is not grasped, because of the abstraction or schematization involved. However, it is not true that there is anything that cannot be grasped by a language system and hence escapes logic. For any single fact in the world, a language system can be constructed which is capable of representing that fact while others are not covered (Carnap 1950, 210).

Apart from the fact that in his argument Carnap reinforces the finite variety assumption, we observe that Carnap binds up intimately the concept of structure with that of a language. Let me call the logical structure generated by a language system *the language embedded structure*. As we have already observed there are extra logical stipulations that in Carnap's view furnish the construction of a language. Hintikka suggests that of these the most important is tying up a language with a given domain.[53]

However, in *Logical Foundations of Probability* Carnap seems to be unclear whether there is only one language embedded structure or more. In the period preceding the publication of *The Continuum of Inductive Methods* (Carnap 1952b), R. Carnap seems to have been certain that there is only one language embedded structure, which corresponds to one language of unified science (Carnap 1938/1949, 415-21). In the above quoted passage, on the other hand, Carnap seems to endorse the idea that, since there is "a very small number of fundamental magnitudes to which ... the great variety of phenomena is ultimately reducible", there is only one language embedded structure, which we are not able to recognize yet. However, the circle analogy and Carnap's commentary that in the case of logic these different methods of covering "facts" are obtained by introducing new predicates as well as the example of geometry seem rather to suggest that the finite variety assumption is local, i.e. it is relative to a given domain of "the great variety of phenomena", and that there is no global language embedded structure as much as there is no global metric (Hacking 1971, 589). Richardson observes a similar ambivalence in the *Aufbau*, due to which Carnap subsequently rejected the whole construction developed in this book (Richardson 1998, 204-208). In the case of there being multiple language embedded structures, which differ with respect to their domains, sets of primitive predicates and their total number *j*, we face the problem of how to chose between them. The suggestion is that since the epistemic canons are built in a language embedded structure, and it therefore makes no sense to compare language embedded structures on epistemic grounds, some other criteria of evaluation must be applied. Since on the grounds of *Logical Foundations of Probability* it is possible to defend the view that there is only one language embedded structure as well as the view that there are more than one, I will not take either side; for practical reasons though I will use the term "language embedded structure" in the singular.

What is illuminating for Carnap's approach to a language embedded structure is his response to Quine's "Two Dogmas" (Quine 1950/1996). He agrees with Quine that "in revising the total system of science no statement and no rule is immune or sacrosanct", and even that "[o]ne can consider replacing the usual form of logic by an intuitionistic one or a three-valued one or whatever". Nevertheless, Carnap disagrees with Quine's conclusion that "there is no sharp boundary between physics and logic", and emphasizes that "[t]he difference between analytic and synthetic is a difference internal to two kinds of statements inside a given language structure; it has nothing to do with the transition from one language to another" (Carnap 1990, 431).

Quine's response is the following:

> Insofar as we are content to think of all science as fitted within that stereotyped logical framework—and there is no hardship in so doing—our notion of logical vocabulary is precise. And so, derivatively, is our notion of logical truth. But only in point of extent. There is no epistemological corollary as to the ground of logical truth (Quine 1962, 59).

And by the same token "there is no epistemological corollary as to the ground" of Carnap favorite derivatives of "logical truth", such as *state-description* or *range*. In leveling this objection, however, Quine mistakenly attributes to Carnap the empiricist conception of logic. To appreciate Carnap's own understanding of what determines a language embedded structure it is necessary to acknowledge that "our empirical *language* can only be understood as an incoherent and fragmentary schema of an ideally coherent *language*" (Sellars 1948/1959, 456).

In sum, we cannot conclude that there are multiple language embedded structures in Carnap's account for they might only be instrumental in developing the final language embedded structure, which will be adequate to the mature form of science. However, Carnap admits here that the structure specification, which yields epistemic canons for inferences, does not turn on a particular form of logic. But even if there are changes as to admissible systems of logic to generate a language embedded structure, it does not imply that the epistemic canons cease to be determined in a purely formal and analytical way, although, as Carnap further admits, we may face severe problems in attempts to apply a given language embedded structure to actual science, which is always still in-the-making, and whose "current degree of explicitness of the rules of the language... is still very small" (Carnap 1990, 432).

Let me come back to the issue of the assessment of Carnap's general epistemological position in *Logical Foundations of Probability*. We have already noticed that all necessary and sufficient conditions for inferential justification beyond those specified by a language embedded structure are not sufficient nor necessary conditions for evaluation of the epistemic status of a statement within a language embedded structure. The former concern only the stipulations of the applicability of the latter to a given knowledge situation. Thus the language embedded structure with respect to its own statements may be identified as a reliable mechanism of epistemic evaluation, because the language embedded structure—for all inferential relations between its own statements—uniquely determines the epistemic canons of evaluation.

The examination of Carnap's epistemological theory of justification carried out in this section allows me to come to grips with the necessary condition of justification propounded on the grounds of structural reliabilism, and outlined in definition 2.10 of the recursive condition (RSR) (see section 3 of this chapter, page 76).

It is crucial at this point to unravel the meaning of the locution "${}^{ST}p_2$ is justified by ${}^{ST}p_1$" in accordance with Carnap's views. As pointed out, for Carnap the pattern of the structural translation of scientific sentences is provided by the history of geometry. The full details of the structural translation for *all* scientific sentences are not worked out by Carnap, because the "current degree of explicitness of the rules of the [scientific] language... is still very small". The design of the structural translation turns also on the unresolved issue of whether there is only one language embedded structure for unified science or more.

If, however, a given language system is specified, then it is possible to construct state-descriptions, and to determine the range of each sentence of this language system. Finally, it is possible to determine the value of the measure function $m(.)$ for each sentence of the language system, and hence the value of $c(.\,|\,.)$ for any pair of sentences of the language system. At this stage, the value of $c(.\,|\,.)$ supervenes on the structure of its arguments and is weighted by parameters reflecting features of the structure of the language system as a whole. Granted this, as a first approximation we should apparently bring out the reliabilist condition for *quantitative* justification:

DEFINITION 2.12 *Reliabilist condition for quantitative justification* (Q)

${}^{\mathrm{ST}}p_2$ *is* justified *by* ${}^{\mathrm{ST}}p_1$ to the extent *r iff* $c({}^{\mathrm{ST}}p_2\,|\,{}^{\mathrm{ST}}p_1) = r$.

There are many ways to transform this rather straightforward quantitative explication of "${}^{\mathrm{ST}}p_2$ is justified by ${}^{\mathrm{ST}}p_1$" as provided in (Q) into a qualitative form which would be appropriate for (RSR). Since I emphasize here the *minimalist* character of the structural reliabilist theory of justification, I propose the following transformation:

DEFINITION 2.13 *Reliabilist condition for minimal justification* (MJ)

${}^{\mathrm{ST}}p_2$ *is* (minimally) justified *by* ${}^{\mathrm{ST}}p_1$ *iff* $c({}^{\mathrm{ST}}p_2\,|\,{}^{\mathrm{ST}}p_1) > \partial$.

Apart from the well-known problems this condition raises, there is one characteristic of the theories of justification. While there are two most plausible candidates to determine the value of ∂ (discussed below), the choice will turn on the kind of the concept of justification: *synchronic* or *diachronic*. Granted the evidence is fixed, the issue of synchronic justification for a given set of sentences can be pursued. However, if evidence varies with time, then the question of diachronic justification of a given sentence would perhaps be more pertinent.

In the context of synchronic justification, where evidence—${}^{\mathrm{ST}}p_1$ in this case—is fixed, the value of ∂ should presumably depend on the degree of confirmation of the hypothesis most directly related to ${}^{\mathrm{ST}}p_2$, i.e. its negation $\neg{}^{\mathrm{ST}}p_2$, as follows:

$$c({}^{\mathrm{ST}}p_2\,|\,{}^{\mathrm{ST}}p_1) > \partial = c(\neg{}^{\mathrm{ST}}p_2\,|\,{}^{\mathrm{ST}}p_1) \qquad \text{(MJS)}$$

which—by addition (proposition 1.10) and normalization (proposition 1.17)—is equivalent to:

$$c({}^{\mathrm{ST}}p_2\,|\,{}^{\mathrm{ST}}p_1) > 1/2 \qquad \text{(MJS)}$$

In the context of diachronic justification, the value of ∂ should reflect the dynamic character of evidence:

$$c({}^{\mathrm{ST}}p_2 \mid {}^{\mathrm{ST}}p_1) > c({}^{\mathrm{ST}}p_2 \mid t) \qquad \text{(MJD)}$$

where *t* stands for tautology.[54]

A prima facie objection against this transformation of the quantitative concept of justification (Q) into the qualitative one is that the minimal justification is not belief-conducive. It is the burden of the next section, however, to discuss the viability of this proposal.

Even if we agree that the minimalist kind of theory propounded by Carnap constitutes in fact a genuine *epistemological* theory, it is still problematic in what sense this theory is a *reliabilist* theory of justification? Admittedly, the reliability concept as applied to Carnap's theory of justification diverges from common epistemological usage in several respects.[55] For instance, initially "reliable" was attributed to processes, which generate a belief, by virtue of the fact that the type to which the former belong displays a sufficiently high ratio of true to false beliefs. Further, each process is realized by an empirical epistemic agent. The earlier discussed repudiation of the relevance of empirical epistemic agents by Carnap entails also the epistemic irrelevance of their psychological processes. For Carnap this is not an actual psychological process which has to be addressed in order to determine the reliability of the resulting belief. For Carnap it is a language embedded structure that carries out the task of determining *quantitatively* the reliability of each sentence which is formed in this language relative to any other sentence formed in the same language. This kind of reliability of a given sentence is determined by its structural properties as stipulated by (SMT), irrespective of the actual formation of a given sentence within a language embedded structure. Therefore it is appropriate to call this kind of minimalist reliabilism *structural* in order to distinguish it from the mentalistic kinds of reliabilism, and from process reliabilism in particular.

A major point of disagreement between minimalist and mentalistic kinds of reliabilism is cogently depicted by W. Alston. In his analysis of the concept of reliability he points out that the weakest and most attacked assumption of reliabilism is that concerning the functioning of the human psyche. Let me briefly outline his analysis. The inferential belief-formation reliable processes are functions which are psychological realizations of the principles of inference, and which take as input knowledge states about a given domain. It is a recognized difficulty among reliabilists that they do not come up with a principled indication on how to determine the relevant type of belief-forming processes which are to be evaluated as reliable or not. Alston claims that each formation of inferential belief involves the psychological activation of a function which determines the content of the output belief on the basis of the contents of input beliefs. What is important, however, is that the semantical specification of the content of the output belief may vary, for it might be specified in a more or less general way, but then its reliability will also vary according to the semantical specification of its content. Alston then emphasizes the fact that what is to be taken into account is the belief actually produced which has a determined level of generality. When the belief is semantically specified in a more general way, then it is no longer a belief produced by that particular process. However, this last claim requires—according to Alston—a realist position on the human psyche, i.e. the human psyche is determined to actually produce beliefs each of which has only one adequate semantical specification. Alston concludes thus: "the viability of a reliabilist theory of justification or knowledge hangs on the viability of psychological realism" (Alston 1995b, 19).

Admittedly, Alston advances this point with respect to the mentalistic concept of reliability. The essential part of his claim, however, can also be formulated with respect to the minimalist concept of reliability. Suppose we write down sentences which represent the propositional content of the output belief and of input beliefs, and suppose further that we manage to structurally translate those sentences. Then it follows that the language in which the sentences are formulated has a language embedded structure, which "completely and exactly" determines the ranges of those sentences, i.e. their semantic properties. We can study the mutual relations between those sentences not only on the grounds of their semantic properties, but also on the grounds of the metric function defined for the language embedded structure. It seems then that in order to specify the "function of content" invoked by Alston it is not necessary to refer to the process of formation of the sentences. In fact, Carnap's structural reliabilism takes the difference between how sentences are generated and what constitutes their "inferential structure" to its extreme; cf. (Audi 1998, 155).

5. The Finnish School's Systems of Inductive Logic and Structural Reliabilism

In section 3 of the first chapter I discussed in detail how far the Finnish School's developments outperform the initial systems of inductive logic proposed by Carnap. A natural question is, then, what, if any, is the *epistemological* outcome of the differences between the two kinds of systems of inductive logic? I argue that the answer is twofold. Remarkably, the structural reliabilist theory of justification seems to be invariant with respect to the modifications introduced by the Finnish School. Nevertheless, those modifications reflect an immense divergence with respect to how Carnap and the Finnish School proceed with integrating the structural reliabilist theory of justification into the theory of knowledge.

Among the differences between Carnap's and the Finnish School's systems of inductive logic, two are pertinent to the discussion of the problem of whether Carnap and the Finnish School embrace the same kind of theory of justification. As described in the third section of the first chapter, the prompt for construing an alternative system of inductive logic for Hintikka was the zero-confirmation problem of generalizations. Irrespective of how serious this problem is taken to be from the perspective of the philosophy of science, it must be taken into account with respect to epistemological analysis of the two kinds of systems.

The other problem, not mentioned yet, is that of the rules of acceptance of scientific sentences, especially of scientific hypotheses. It was given different solutions by Carnap and the Finnish School.[56] Both Carnap and the members of the Finnish School, however, acknowledge the possibility of the alternative approach:

> Induction can be characterized in two ways: it can be regarded as a process which prescribes or recommends the acceptance of new beliefs and expectations on the basis of logically inconclusive evidence, or induction can be regarded as a method of assigning probabilities or degrees of credibility to statements and hypotheses relative to evidential propositions, without assuming that the hypotheses in question should be "accepted" on the basis of evidence (Hilpinen 1999, 4).[57]

Carnap claimed that there are no rules of acceptance, or at least that none are necessary in science. Hintikka and his co-workers, on the other hand, sustained the view that—especially from the epistemological perspective—such rules are indispensable for science, and this view motivated their account of the acceptance rules on the grounds of the hitherto elaborated systems of inductive logic. Thus, this issue must be considered as one that—purportedly, at least—might have resulted in constructing different theories of justification by Carnap, on the one hand, and the Finnish School, on the other.

As we have observed in the first chapter, Hintikka's strategy in developing the α-λ-system was, crudely, to stick as closely as possible to Carnap's methods in construing the λ-system, with the restriction, though, that generalizations have non-zero confirmation. Both, the domain of the confirmation function and the confirmation function itself are defined—*mutatis mutandis*—in Carnapian possible-world semantics. Accordingly, a language embedded structure is generated, which establishes the set of epistemic canons against which to assess the epistemic status of "inductively inferred" sentences. What might, however, countenance the view that it is again the logico-mathematical structure which is necessary and sufficient to determine the epistemic status of a proposition is Hintikka's pronouncement that the value of α is to be set up synthetically, i.e. not by logical means alone. This claim is basically grounded in the *presupposition view of induction* evoked by Hintikka; it finds further support—as will become clear below—in the way Hintikka and his co-workers set up acceptance rules.[58]

On the basis of how the parameter α influences the confirmation values of generalizations in the α-λ-system of inductive logic, Hintikka concludes that α is an *index of caution*, and that it reflects one's convictions about the kind of order which one assumes to be out there in the world.[59] M. Mondadori assents: "[T]he choice of an assignment of values to these parameters represents far-reaching *factual* assumptions about the *objective* structure of our field of inquiry" (Mondadori 1987, 176). The problem, however, is that neither of these interpretations can be directly represented in any of the Carnap-Hintikka style systems of inductive logic. Thus, it seems that Carnap is right in claiming that the question of the interpretation of the parameters of the systems of inductive logic is not a matter of inductive logic alone, but it is a matter of *methodology*, i.e. the application of inductive logic to a particular problem-situation. This claim gains further support from Zabell's modification of the sufficientness postulate, which leads from Carnap's initial γ-λ-system to the minimally modified system with non-zero confirmation of generalizations (Zabell 1997). The point is, then, that the modification required by the systems of inductive logic that involve the zero-confirmation problem is not a matter of *principle*, i.e. it seems to fit the initial epistemological framework set up by Carnap and analyzed in the previous section.

Let me now turn to the issue of acceptance rules.[60] In both probability theory and in epistemology the most common acceptance rule is *the rule of high probability*:[61]

DEFINITION 2.14 *The rule of acceptance based on high probability* (AHP)

$$Ac(h \mid e) \text{ only if } P(h \mid e) > 1 - \varepsilon;$$

where $0 < \varepsilon \leq 0{,}5$.

Lehrer acknowledges that the rule of high probability (AHP) is only one of the possible strategies of acceptance. There might be also extreme strategies: if someone aims at acceptance of *all* true sentences, or *no* false sentences, then a more appropriate rule would be to accept every sentence, or, accordingly, to accept no sentence (Lehrer 1981a, 49).

If, in addition to the rule of high probability (AHP), we introduce two further requirements: that the set of accepted sentences be closed under logical consequence and consistent, then the lottery paradox arises.[62] Not surprisingly, there are, in general, two approaches with respect to the rule of high probability.[63] On the one hand, not only is the rule of high probability rejected, but also any possibility of construing such rules is denied. Further, it is claimed that the rule of acceptance is only mistakenly attributed an indispensable role in the course of science, and this role is overtaken by some other procedure. The acceptance rules, on the other hand, are supposed to be the most legitimate procedures in science, and it is merely their appropriate form that is debatable. Acceptance rules functioning as a counterpart to the rule of detachment of deductive logic not only defend the status of inductive inferences as such, but also seem indispensable for scientific explanation (Kyburg 1968b, 99).

Hintikka and Hilpinen embrace the view that acceptance rules are indispensable for science and seek to modify the rule of high probability.[64] They implement the solution within Hintikka's α-λ-system and divide the problem into two separate acceptance-problems (Hintikka and Hilpinen 1966, 6); (Hilpinen 1968, 144-45). The first problem concerns the acceptance rule for generalizations, the second for singular sentences. In the case of general sentences the α-λ-system quite naturally leads to the following modification of the rule of high probability:[65]

DEFINITION 2.15 *Generalized* (AHP) *rule for the α-λ-system* (AHG)

$$Ac(h \mid e) \text{ only if } P(h \mid e) > 1 - \varepsilon \text{ and } n > n_0;$$

where n_0 is the threshold value of the number of observations required, which depends upon α (Hintikka and Hilpinen 1966, 14); (Hintikka and Hilpinen 1971, 303).[66]

In practice only those generalizations which in their normal form contain the constituent C_c are themselves acceptable (Michalos 1971a, 299). The rule for acceptance of singular sentences is constructed on the basis of (AHG): only those singular sentences which are the consequence of acceptable generalizations are acceptable (Hintikka and Hilpinen 1966, 15-17); (Hilpinen 1968, 68-74). The fact that only those singular sentences which are derivatives of acceptable generalizations are acceptable is interpreted by the members of the Finnish School as consonant with Aristotle's claim that knowledge concerns only the general (Hilpinen 1968, 76); (Hintikka and Hilpinen 1966, 19; 1971, 306).

Notwithstanding the way Hintikka and Hilpinen define the acceptance rules, it in no way affects the semantical construction of the α-λ-system of inductive logic.[67] To authorize inductive inferences as legitimate kinds of *inferences* Kyburg proposes that acceptance rules play the role of the rule of detachment for inductive inferences (Kyburg 1968b; 99). However, Suppes's clear delineation between probabilistic rules of detachment and acceptance seems to seriously undermine Kyburg's claim (Suppes 1968, 129). Therefore, the epistemic canons of justification specified within the system constitute the necessary conditions for acceptance. Hintikka's own procedure is depicted by the positive answer, he elaborates in greater detail, to the following question:

> [C]an one find probability measures on the statements of some simple language and define by their means a probabilistic rule of acceptance which satisfies the minimum requirements of consistency and logical closure which every well-behaved rule of acceptance clearly should satisfy? (Hintikka 1968d, 144-45).

We have already observed that inductive logic constitutes, as in the case of Carnap, the structural reliabilist theory of justification. It turns out, then, that a sentence is acceptable, according to Hintikka and Hilpinen, only if it fulfills the condition for justification specified by the structural reliabilist theory of justification. Bar-Hillel is right in pointing out the pragmatic character of the concept of acceptance:

> which is not worthwhile, or possible, adequately to explicate as a purely semantic term (Bar-Hillel 1968c, 154-55).

In the view of Bar-Hillel's claim, it is perplexing what precisely Hintikka and Hilpinen have in mind when they write:

> We are not going to discuss the importance of the rules of acceptance from the point of view of the methodology of science. Instead, we shall consider the possibility of incorporating a probabilistic rule of acceptance into a system of quantitative inductive logic, because such a rule would in our view be very helpful for the purpose of explicating in terms of inductive probabilities such classificatory expressions used by most philosophers as "empirical knowledge", "practical certainty" and "rational belief" (Hintikka and Hilpinen 1966, 3).

Since the rule (AHG) depends upon α, it can only be specified in terms of what Carnap calls the "methodology" of inductive logic. So it cannot be this point that Hintikka and Hilpinen are addressing here. They initially motivate the importance of acceptance rules by invoking Chisholm's definition of knowledge. And the passage quoted above clearly indicates that they insist on the fact that the system of inductive logic is sufficient to generate a quite cohesive epistemological position. They emphasize the fact that in the epistemological view of knowledge they develop on the grounds of inductive logic the indispensable part is formed by the rules of acceptance. In other words, they stress the divergence of their epistemological program from Carnap's probabilistic leaning in epistemology.

Carnap being aware of this fact never committed himself to the view that the concept of the degree of confirmation is an explication of the concept of acceptability (Niiniluoto 1973, 419). Thus, what is also required is an auxiliary motivation, which in the case of the Finnish School is established on epistemological grounds. Hilpinen in (Hilpinen 1968, 7-9) advances the following argument: in the case of deductive logic a sentence *h* is *known* only if it is entailed by a true sentence *e*, i.e. it is deductively justified on the ground of *e*. By analogy, in the case of inductive inference, *h* is *known* if it is acceptable on the grounds of *e*, i.e. *h* is inductively justified only if it gets a high confirmation value on *e*. The latter proposal is further appended with the "non-probabilistic" requirement on the number of minimal observations. If these auxiliary conditions are satisfied, as is demonstrably true with respect to the rule (AHG) (Hilpinen 1968, 63) , then this rule: "also provides a consistent definition of knowledge (as far as general hypotheses are concerned)" (Hilpinen 1968, 64).

It seems inadequate, however, to claim that the issue of the acceptance rules does not make *any* difference with respect to Carnap's and the Finnish School's epistemological positions. On the contrary, it accounts for a vast difference. But—as I shall emphasize—the difference does not concern the theory of *justification*, but the theory of *knowledge,* which they built upon the structural reliabilist theory of justification. And it seems apparent that the difference in the attitude towards the acceptance rules is in both cases primarily motivated by epistemological concerns.[68] What are these then? Both Carnap and Hintikka apparently admit that the concept of knowledge must be defined in terms of propositional attitudes, and in particular in terms of beliefs. The difference, however, concerns an account of the nature of beliefs and the problem of how to define the conditions of justification for beliefs in terms of the structural reliabilist condition for justification of sentences. Carnap quite straightforwardly admits that:

> Suppose that inductive logic supplies a simple result of the form "$c(h \mid e) = r$"... If *e* expresses the total knowledge of *X* at the time *t*, that is to say, his total knowledge of the results of observations, then *X* is justified at this time to believe *h* to the degree *r*, and hence to bet on *h* with a betting quotient not higher than *r* (Carnap 1950, 211).

It is not my present task to elucidate Carnap's theory of knowledge. What I find instructive in this locution, however, is that on the one hand, Carnap identifies a quite natural way to proceed from the minimalist theory of justification to the theory of knowledge. On the other hand, the important observation is that he seeks to determine the conditions for knowledge in terms of a *gradual* concept of belief. He does not hesitate to admit that: "the end result of inductive reasoning is not the acceptance of a hypothesis, but rather the assignment of a probability to a hypothesis" (Carnap 1968b, 146).

A radical elaboration of an epistemological position, in which the epistemic status is almost exclusively attributed to gradual beliefs, is due to Jeffrey. In (Jeffrey 1985/1992, 68) he acknowledges that the probabilistic epistemology was motivated by his anti-foundationalism, in opposition to the 17th century foundationalism of anti-probabilists and to the 20th century foundationalism of empiricists like J. Keynes or C. I. Lewis. The latter words his foundationalism succinctly as follows:

> If anything is to be probable, then something must be certain. The data which eventually support a genuine probability, must themselves be certainties. We do have such absolute certainties in the sense data initiating belief and in those passages of experience which later may confirm it (Lewis 1946, 186).

Jeffrey's probabilistic epistemology is much welcomed by Carnap (Carnap 1968b, 146-47; 1971c, 29). The idea of uncertain evidence is consonant with Carnap's views on Protokolsätze (Kistler 1995, 16; Lehrer 1980, 133). In fact, probabilistic epistemology is a descendant of Carnap's epistemological approach: "Carnap's idea of an 'ignorance' prior cumulatively modified by growth of one's sentential data base is replaced by a pragmatical view of priors as carriers of current judgment, and of rational updating in the light of experience" (Jeffrey 1991/1992, 12).Moreover, Carnap's repudiation of the claim that his epistemology is foundational parallels Jeffrey's argument:

> When we speak about the determination of the probability ... we take the evidence as "given"; we treat it as if we knew it with perfect certainty; and this is clearly the same taking it as "accepted" ... And we do this although most of us, when we discuss not inductive logic but epistemology, declare that we can never have perfect certainty for any synthetic proposition. For many years I have tried in vain to ... find a way of dealing with uncertain evidence (Carnap 1968b, 146).[69]

A similar point is also made, but not emphasized by Hintikka:

> all observational reports that are used in testing hypotheses are assumed to be true. Our simple language L_k forces us here into a considerable oversimplification ... This simplification is not perhaps too serious, however" (Hintikka and Hilpinen 1966, 15).

To be sure, Carnap's own epistemological position does not exactly match Jeffrey's program of probabilistic epistemology, but this may be in large part due to technical limitations that Carnap was not able to overcome.

The Finnish School, however, seems strictly confined to the traditional epistemological approach to the theory of knowledge; in particular, they seem to strive for a kind of Chisholmian foundationalism. However, in place of Chisholm's non-gradual epistemic appraisal they propose quantitative measure of the epistemic status of beliefs:

> According to a widespread philosophical view, most of our knowledge of empirical matters of fact is logically speaking not absolutely certain, but only probable. ... It thus lies close at hand to try to define knowledge in terms of truth and high probability, and many philosophers have in fact tried to do so. ... [W]e shall consider some of the difficulties encountered by these attempts, and propose a partial solution to them (Hintikka and Hilpinen 1966, 1).

This is also how they motivate the investigation of the adequate definition of acceptance rules (Hintikka and Hilpinen 1966, 1; Hilpinen 1968, 27-28). First, the Chisholmian condition for knowledge is introduced.[70] Then, it is recast in the form of a minimalist condition as follows:

$$K(h \mid e) = Ac(h \mid e) \text{ and } h \text{ is true} \qquad \text{(DK);}$$

where $K(h \mid e)$ is "the knowability" of h conditional on e (Hilpinen 1968, 28).

The rule (AHG) in some respect constitutes a new paradigm for epistemological investigations concerning the theory of knowledge. For it splits the identity between the concept of acceptance[71] and belief,[72] which, apparently, is embraced by Carnap and by most contemporary epistemologists. If belief is thought of in terms of degrees of confidence, and something we cannot come up with at will (Alston 1989), then the rule (AHG) implies quite a different kind of concept. Nevertheless, this kind of concept is not explicitly developed by the Finnish School.

In his work on acceptance rules[73] Hilpinen was able to trace some links between Carnap's epistemological attitude to the concept of knowledge and the respective attitude of the Finnish School. In order to make the systems of inductive logic applicable without recourse to acceptance rules Carnap introduces "methodological rules for the use of confirmation statements" such as the requirement of total evidence or the rule of maximizing expected utility (Salmon 1968a, 39). This approach, according to Hilpinen's results, is tenable, but what is further required is a suitable definition of utility as well as appending the set of acceptance rules with "the rule of ties", i.e. the rule of suspending judgment with respect to equally acceptable hypotheses (Hilpinen 1972, 167-68).[74]

Moreover, Carnap seems to have changed his own views concerning the possibility of acceptance rules in a way, which allows for an elaboration of the view consonant with the Finnish School.[75] The persistent difficulty, however, is that despite the fact that Carnap has no objections *in principle* to there being rules of acceptance, he refuses to use the concept of acceptance as a non-quantitative concept. For Carnap the distinction between quantitative and non-quantitative concepts is epistemologically substantive, and may constitute a serious obstacle with respect to the project I am here delineating (Carnap 1968b, 147-49). For instance, he suggests that the qualitative concept of acceptance has to be replaced by its quantitative counterpart:

> The function of inductive logic begins *after* a hypothesis is offered for examination. Its task is to measure the support which the given evidence supplies for the tentatively assumed hypothesis (Carnap 1953, 195).

As the study of the theory of knowledge advanced by Carnap and the Finnish School extends beyond the scope of this book, I will not pursue this discussion any further. What is still left for investigation, to be undertaken in the next chapter, is the evaluation of the epistemological import of the structural reliabilist theory of justification explicated in the systems of inductive logic.

Notes

1 The requirement of psychological plausibility is naturally imposed on mentalistic theories, because of (QT)—the requirement to be discussed in the ensuing.

2 D. T. Campbell advocates a theory of justification, where the justification-transmission is primarily carried out by social interactions; cf. (Campbell 1986; 114-15). A different argument is advanced by S. Fuller, who claims that it is not possible to disentangle clearly between the individual subject and the social subject of knowledge; cf. (Fuller 1989, 17), and (Tuomela 1992) for a similar point.

3 For a criticism against presupposing logical omniscience on the part of epistemic agents see e.g. (Goldman 1986, 279-80). E. Adams presents two arguments, which point out—with respect to geometry in the first instance – that by no means can we develop a theory by simply imposing more and more constraining idealizations upon our ordinary beliefs; cf. (Adams 1961, 198). Contrary to that, Carnap claims that "Euclid's axiom systems of geometry was a rational reconstruction of the beliefs concerning spatial relations" (Carnap 1945/1970, 472).

4 For a discussion of the relevance of psychological experiments on deductive and probabilistic reasoning—in particular the fallacies in reasoning process—to epistemology see e.g. (Schlick 1935/1949, 404); (Tversky and Kahneman 1977); (Sober 1978); (Cohen 1979); (Kahneman et al. (eds.) 1982); (Tversky and Kahneman 1983); (Goldman 1986, 305-23); (Klayman and Ha, 1987); (Kyburg 1987); (Musgrave 1989); (Kornblith 1993);

(Kusch 1995, 30-62); (Fisak and Pidgeon 1997, 1-3); (Picardi 1997); (Braine and O'Brien (ed.) 1998, chapters 2 and 3); (Stanovich and West 2001).

5 Except for what E. Sosa terms as *formal foundationalism*, i.e. the thesis that "[a]ll epistemic justification, authority, warrant, or any other epistemic status of belief hence derive from what is not epistemically evaluative. Any epistemically justified belief must have non-epistemic properties which make it thus justified, so that any other belief sharing all such properties would be equally epistemically justified" (Sosa 1991, 110).

6 For a critical and historical account of deontologism prevailing in contemporary theories of justification see (Plantinga 1993; 1995a); (Bronk 1997, 9-19). For a discussion of how epistemic obligations arise on the grounds of a minimalist theory of justification see (Kyburg 1980, 93-94).

7 As W. Alston admits it is characteristically reliabilism, however, to constitute "the most common way of further specifying externalism" (Alston 1998, 821); see also (Goldman 1998, 206).

8 For the most comprehensive account of the types of foundationalism see (Bronk 1998b, 145-78). See also (Pollock 1986, 26-65).

9 These are, indeed, two basic epistemological alternatives one shall start with; cf. (Lehrer 1974, 76); (Pollock 1986, 66-92); (Plantinga 1993, 66-86); (BonJour 1998b, 253).

10 The difference between mentalistic and minimalist theories of justification turns on the account of belief. For instance, in K. Lehrer's epistemology belief plays only an instrumental role of selecting a given set of propositions, while the necessary and sufficient conditions for justification are specified in non-epistemic terms; cf. (Lehrer 1981a, 75). Similarly a role of belief is diminished on R. Wójcicki's account; cf. (Wójcicki 1995/1996a, 354).

11 A. I. Goldman's position in the 1979 paper is called "process reliabilism". However, he subsequently modified his position; cf. (Goldman 1986; 1993). It is commonly recognized that the first version of reliabilism was formulated by F. P. Ramsey in (Ramsey 1931), who claimed that for a belief to constitute knowledge is to be true, certain and to result from a reliable process. Other formulations of reliabilism preceding A. I. Goldman's famous paper include: F. Dretske (Dretske 1971a); D. M. Armstrong (Armstrong 1973); and concurrent with Goldman: R. Nozick (Nozick 1981) and M. Swain (Swain 1981). For an account of Goldman's changing views see (Sosa 1991, 132-5); (Goldman 1993; 1998). For an excellent discussion of different forms of reliabilism see (Goldman 1986, 43-53); (Pollock 1986, 114-16); (Plantinga 1995c, 427-34); (Goldman 1998, 207-208). For a response to Plantinga's objections see (Alston 1995a, 398-402). For a criticism of Goldman's process reliabilism see e.g. (Ring 1986); (Almeder and Hogg 1989); (Sturgeon 1997, 131-45).

12 Goldman's reliabilism, in addition to the issues discussed in this chapter, raises an important issue of naturalistic epistemological theories. R. Almeder claims that Goldman in (Goldman 1986) advances a sophisticated version of normative naturalistic epistemology; cf. (Almeder 1990/1993, 451). Almeder's point, I believe, will be successfully met on the grounds of a minimalist theory of justification, which I will introduce in the ensuing.

13 In his detailed analysis in (Alston 1995b) Alston comes to the compelling conclusion that any mentalistic reliabilism must assume psychological realism; otherwise the position is untenable, because of the *Generality Problem* (see page 71) and *No-distinction Problem*; cf. (Feldman 1985, 159-60); (Goldman 1986, 49).

14 Goldman does not use the word "quietistic", but still the description of his own theory accords well with (QT); cf. (Goldman 1979/1993, 89).
15 In fact, the problem is characteristically strenuous in the case of conflicting norms of and actual instances inference; cf. (Stich and Nisbett 1980/1997, 174-82).
16 For arguments against including "truth" among epistemic concepts see (Goldman 1986, 144-51).
17 For a discussion of the difference between epistemic and non-epistemic concepts see also (Coffa 1974, 148-50).
18 Since Goldman is silent on this point, it seems that he takes (QT) to be compatible with (ST). This, however, will only be true in the case where the "ordinary standards" of evaluation of belief's epistemic status do not involve epistemic principles. B. Stroud and E. Nisbett by referring to psychological experiments on inductive fallacies demonstrate, however, that A. I. Goldman's theory of justification cannot embrace (QT); cf. (Stroud and Nisbett 1980/1997, 274-82).
19 Goldman is well aware of this problem, see (Goldman 1979/1993, 97). For an extended account of the problem see (Feldman 1985); (Sosa 1991, 131); (Alston 1995b).
20 Admittedly, a theory of direct justification to complement structural reliabilist theory of justification explicated in this book is not straightforward, either. Presumably, the most plausible candidates would be picked out from externalist theories.
21 For an outline and discussion of Doob's theorem as well as Gaifman and Snir result see (Earman 1992, 144-49).
22 The pertinent alternative is structural realism; cf. (Worrall 1989); (Psillos 1996); (Ladyman 1998).
23 For a critical analysis of Carnap's motivation to apply epistemology exclusively to scientific sentences see (Toulmin 1966, 88-89); (Bronk 1995, 202).
24 For an examination of the theory of knowledge projected by the members of the Finnish School see e.g. (Hilpinen 1977).
25 Cf. (Plantinga 1993, 114-37); (Pollock 1986, 92-102). See (Horwich 1993, 41) for an opposite view.
26 A. Burks, for instance, advances precisely this claim, namely, that Carnap's system of inductive logic is a purely formal construction, without any relevance to epistemological issues; cf. (Burks 1963; 739).
27 According to R. C. Jeffrey de Finetti was the first to embrace probabilistic epistemology in (De Finetti 1931/1989); for an elaborated exposition see also (Jeffrey 1956; 1968a; 1968b; 1992); (Eells 1994); (Kaplan 1996); (Skyrms 1996b); (Howson 1997a, 270-79); for a survey see (Pollock 1986, 105-14); (Galavotti 1997); (Loewer 1998). See (Hosack 1991) for a critical assessment.
28 The fact that Carnap's logic, in particular inductive logic, is an epistemological theory is readily acknowledged by contemporary epistemologists, cf. e.g. (Chisholm 1966/1977, 71); (Goldman 1986, 6-7).
29 Notwithstanding later accomplishments of the systems of inductive logic by Carnap, Hintikka and others, the substantial motivation for the epistemological position espoused in the systems of inductive logic is apparently brought out in *Logical Foundations of Probability*.
30 Cf. (Richardson 1998, 144). Carnap's logicism has not only evolved, but is also essentially different form G. Frege's and B. Russell's; cf. (Bohnert, 1975a); (Thiel 1982), and from

C. G. Hempel's; cf. (Hempel 1945/1949b, 225-33). For an argument against tenability of logicism in general see e.g. (Thom 1982, 5).

31 Indeed, in Carnap's logicism numbers are purely logical as individual constants; cf. (Carnap 1930/1959); (Bohnert 1975a, 208).

32 For a response to I. Lakatos's criticism see (Adler 1980, 382-86).

33 For an account of foundationalism in terms of the structure of justification see (Bronk 1998b, 152-53).

34 On this broad characterization of foundationalism, it does not preclude coherentism; cf. (Lehrer 1981b, 234). Nevertheless, foundationalism is here a substantive, and not merely formal theory—for the distinction see (Sosa 1991, 154; 1998, 718-19).

35 E. Sosa embraces a similar view; cf. (Sosa 1991, 24).

36 For a definition of coherence alternative to the Bayesian one see (Lehrer 1981a, 82-88). L. BonJour discusses difficulties underlying the definition of coherence as well as attempts to overcome them; cf. e.g. (BonJour 1985; 1998b, 253-56).

37 My explanation of "coherence" and "strict coherence" follows Bayesian exposition thereof, but my discussion does not turn on the Bayesian reading of these concepts. For Carnap's original exposition of strict coherence see regularity conditions discussed in chapter 1, page 17.

38 For pertinent discussions of coherentism see e.g. (BonJour 1985); (Lehrer 1974; 1990).

39 E. Sosa's *self-abstraction* and *present-abstraction* arguments draw precisely on this point; cf. (Sosa 1991, 76-78).

40 Though it might, for instance, explain why Carnap was perceived as a coherentist by Schlick (Oberdan 1998, 299); see also (Schlick 1934/1959); (Hilpinen 1982); (Koppelberg 1998, 256-59). In particular, the version of K. Lehrer's coherentism exposed in (Lehrer 1974) is definitely misleading from Carnap's perspective as it ignores conditioning on evidence, which results in "beliefs [that] may be completely justified even though beliefs essential to the justification are false" (Lehrer 1974, 214).

41 For a survey of alternative concepts of knowledge which do not entail belief see (Luper 1998, 706).

42 In particular the logical relation between evidence and *S*'s belief is not sufficient to yield the evidential support; cf. (Feldman 1993, 121).

43 It seems that one of Carnap's predecessors in construing a highly idealized theory of justification was W. Sellars; cf. (Lehrer 1974, 45).

44 For a defense of a similar view see (Chatalian 1952).

45 For a similar view see (Niiniluoto 1972).

46 In R. Audi's terminology R. Carnap embraces thereby *the entailment principle*; cf. (Audi 1998, 300).

47 Because of the eminent epistemic status of the meta-level sentences, it is appealing that Carnap attributes logical form with a kind of justificatory holism. For an argument see (Sosa 1991, 280).

48 See (Kyburg 1960/1961); (Maher 1998, 757-58) for a historical survey of similar positions, and (Kyburg 1968a) for a discussion of Carnap's arguments.

49 R. Chisholm investigates a number of problems that (MR) ends up with in terms of "application fallacies"; cf. (Chisholm 1982, 34-39). F. I. Dretske, on the other hand, poses a paradox which seems inevitable if in defining the concept of knowledge we draw upon the probabilistic concept of evidence; cf. (Dretske 1971). However, it is another paradox, the so called paradox of "old evidence", that was widely attended to by probabilistic

epistemologists. It is paradoxical that on Bayes's theorem the evidence available prior to performing an experiment does not increase the probability of a scrutinized hypothesis; cf. (Jeffrey 1983); (Skyrms 1983); (Howson 1991b); (Jeffrey 1991/1992, 103-107); (Earman and Salmon 1992, 98-99); (Jeffrey 1995); (Wagner 1997).

50 A similar procedure is traced by A. Richardson in the *Aufbau;* cf. (Richardson 1998, 175).

51 Coffa suggests that already in the early 1930s Carnap came to this conclusion as regards all foundational studies and all of epistemology; cf. (Coffa 1993, 309).

52 This seems a further sophistication of the doctrine of the relativized a priori adopted by Carnap from neo-Kantians; cf. (Richardson 1998, 196).

53 Cf. (Hintikka 1991a, 328). For Carnap's best exposition on the role of language in solving the putative philosophical problems see (Carnap 1950/1956, 207-208).

54 On the grounds of the systems of inductive logic (MJD) is complemented with strict coherence, which in this case would require that for any ${}^{ST}p_2$ other than logically false sentence, i.e. counter-tautology, $c({}^{ST}p_2 \mid t) > 0$.

55 For a general definition of reliabilism see (Sosa 1991, 131).

56 See (Michalos 1971b, 3-15); (Batens 1975, 280-98) for a survey.

57 See (Hempel 1981) for a similar view. See (Kyburg 1970, 61-66; 1976, 192-99) for a critical discussion.

58 A point I do not pursue here concerns the problem of the consistency of the "synthetic" interpretation of the α-parameter on the grounds of inductive logic *alone*; in (Kawalec 1998) I identify some problems that seem to undermine this approach.

59 See also (Hintikka 1968a, 303-306); (Pietarinen 1972, 109-111; 127-131); (Hilpinen 1971, 281-86).

60 For a comprehensive overview see (Salmon 1968c); (Levi 1976); (Hajduk 1984, 74-97; 168-82).

61 Cf. e.g. (Chisholm 1957, 27-28); (Hempel 1962, 155); (Kyburg 1961, 1965b); (Sellars 1964, 220-21).

62 Cf. (Kyburg 1961, 71); (Kyburg 1965b, 301-10); (Hintikka and Hilpinen 1966, 4); (Hilpinen 1968, 39-42). H. Heidelberger argues that this paradox constitutes sufficient ground to dispense with probabilistic epistemology; cf. (Heidelberger 1963).

63 I only depict here rough approximations of the positions in the discussions, which are numerous. For instance, some would undermine the conditions imposed on knowledge, such as logical closure; cf. (Kyburg 1968b; 118).

64 For a critical discussion see (Hilpinen 1968, 66-67); (Michalos 1971b, Ch. 8); (Michalos 1971a); and (Hintikka and Hilpinen 1971); (Niiniluoto 1973, 428-32) for a response.

65 (Hintikka and Hilpinen 1966, 11). The rule proposed by Hintikka and Hilpinen is further subjected to a fine generalization and sophistication in the work of Mortimer; cf. (Mortimer 1982, 10.2).

66 A. Michalos in (Michalos 1971a, 299) points out that n_0 inherits the problems of non-arbitrary choice that threaten the α-parameter. For some other problems see (Schick 1970, 10-14); (Kawalec 1998).

67 This fact is further supported by other approaches to the rules of acceptance, notably Kyburg's; cf. (Kyburg 1968b; 1980); (Salmon 1968a, 75).

68 Salmon advocates the view that Carnap evades the acceptance rules since they can lead to inconsistencies (Salmon 1968b, 38). Being a technical point, this respect is of minor importance for Carnap.

69 For an idea and elaboration of "uncertain evidence" see (Jeffrey 1965, 153-70; 1987/1992); (Niiniluoto 1997, 123-28). For a criticism of Carnap-Jeffrey approach to acceptance rules see (Levi 1976, 11-14).

70 "'X knows that *h* is true' means:

(i) X accepts the hypothesis *h*,

(ii) X has adequate evidence for *h*, and

(iii) *h* is true;"

(Chisholm 1957, 16). I quote from (Hilpinen 1968, 27). See also (Hintikka and Hilpinen 1966, f.1 on p. 1).

71 On the misleading use of the concept of acceptance in the discussions on acceptance rules and on various possible meanings thereof see (Bar-Hillel 1968b, 124); (Hajduk 1984).

72 This approach was common to such divergent currents in Polish epistemology as R. Ingarden's epistemology, on the one hand, and K. Twardowski's, on the other; cf. (Kawalec 1997a; 1999). For an investigation of the historical explanation; see (Kawalec 1999). For recent elaborations on the distinction see (Lehrer 1981a, 79-80); (Cohen 1993, 130-33); (Kaplan 1996, 102-54); (Roorda 1997, 127-31).

73 Cf. (Hilpinen 1968, ch. 9); (Hilpinen 1972). See also (Hintikka and Pietarinen 1966).

74 See also (Hilpinen 1968, Ch. 9); (Niiniluoto 1987, Ch. 12); (Festa forthcoming, 3.3.2).

75 Carnap in (Carnap 1963, 972-73) seems to suggest that he does not object *in principle* to there being acceptance rules for generalizations; cf. also (Kyburg 1968, 119). An opposite view on Carnap with this respect is defended by Y. Bar-Hillel in (Bar-Hillel 1968b, 122-23). For a critique of Carnap's views see (Hilpinen 1968, 31-35).

Chapter 3

STRUCTURAL RELIABILISM AND THE PROBLEMS OF BAYESIAN EPISTEMOLOGY

The Bayesian reading of the Carnap-Hintikka style systems of inductive logic discussed in section 4 of the first chapter purportedly gives reason to claim that Carnap and Hintikka advance their own peculiar Bayesianism (Plantinga 1993, 114). As it appears, however, both the critics and the proponents of inductive probability apparently fail to appreciate Carnap's and Hintikka's *own* grasp of Bayesianism. I claim that the appropriate way to elucidate the specific version of Bayesianism propounded by Carnap and Hintikka is via the epistemology underlying the systems of inductive logic and in particular, via the structural reliabilist theory of justification. I claim that it is structural reliabilism that grounds a better comprehension of Carnap's and Hintikka's place in the central division of Bayesians into objective and subjective. It is not my purpose, however, to give more credit to Carnap's Bayesianism than it actually deserves, and to defend it against all criticism; rather I develop a reading of Carnap's and Hintikka's Bayesianism that could be shared by both the pros and the cons.

I proceed to argue in this chapter as follows. In the first section, I outline the standard view on Carnap's Bayesianism, and point out that it motivates the most devastating criticisms of the program of inductive logic in general, and the underlying epistemology, in particular. However, this devastating character of criticisms is due to a mistaken view that Carnap's Bayesianism is generically the same as the standard Bayesianism. The question arises whether and how Carnap's own version of Bayesianism draws upon his structural reliabilist theory of justification. This way of motivating Carnap's structural reliabilism is alternative to the one presented in the second chapter of this book where the discussion of an apparent incoherence among Goldman's meta-level principles of his own theory of justification leads naturally to a modified framework for reliabilist theories of justification, a sub-species of which is Carnap's structural reliabilism.

Thus, in the remaining sections I will try to elucidate the rationale for advocating the structural reliabilist theory of justification. I will begin the discussion with those problems of other sorts of reliabilism[1] that only structural reliabilism seems to be able to cope with successfully. There are also epistemological objections against the whole program of confirmation theory, and I shall discuss them insofar as they may appear to threaten structural reliabilism. Apart from the fact that structural reliabilism attains a minimalist theory, it also seems to offer an unmatched alternative to the solution of the Bayesian objective-subjective dichotomy concerning the conditions which determine prior distributions, as I argue in the last section.

1. Epistemological Criticisms of Carnap's Inductive Logic

Around the middle of 1963 Carnap wrote in a letter to de Finetti:

> As I see the situation, the differences between the various conceptions of probability which belong to the second class [normative theories] are not very essential. In particular, the distinction between your and Savage's conception on the one hand, and my conception on the other, seems to me to consist essentially *that I state more requirements than you* (Constantini 1989, 387).[2]

Let me call this *the Bayesian-cum-extra* view on inductive logic. While the Bayesian part consists of normalization,[3] non-negativity, and finite additivity (see page 7), the "cum-extra" part containing the additional requirements, includes the regularity and symmetry conditions (exchangeability) and convergence (see pages 17–20). It is quite unnecessary to address at this point the issue whether a decade earlier Carnap himself appreciated the Bayesian-cum-extra view—in fact, in the period from the early '40s till 1962, which encompasses Carnap's two major works: *Logical Foundations of Probability,* and *The Continuum of Inductive Methods,* Carnap was far from endorsing this view. What is relevant to the present discussion though is the fact that a couple of decades later his critics and proponents still endorse this assumption.

It appears that embracing the Bayesian-cum-extra view on inductive logic is even prima facie inadequate for the purpose at hand. For de Finetti insisted upon finite additivity, while the system of inductive logic allows, in principle, countable additivity. However, the conditions imposed on the construction of the language, which will be discussed in more detail in the next section, endorse finite additivity in practice.[4] It is also far from clear how the Bayesian-cum-extra view, which prima facie admitted of there being empirically determinate frequencies, will fit de Finetti's radical subjectivism. In de Finetti's view referring to "objective chances" or "physical frequencies" would be pointless; there are only respective degrees of beliefs. Carnap, however, was eager to admit that his theory of "the logical concept of probability" is perfectly compatible with Reichenbach's concept of probability as relative frequency, i.e. with an empirical concept of probability (Carnap 1950; 1952); (Carnap and Stegmüller, 1959).

One of the problems with the Bayesian-cum-extra view might be that it is biased towards the classical *personalist* Bayesianism which is committed to the view that the standard axioms of probability theory exclusively constitute the necessary and sufficient conditions for a theory of confirmation, and any other axioms would be superfluous. The problem I find more pressing rather is the piecemeal epistemological approach endorsed by the Bayesian-cum-extra view. For it assumes either that it is epistemically neutral to add extra axioms to the standard axioms of probability theory or that each extra axiom added to the standard axioms of probability theory has easily tractable epistemological consequences. That both assumptions are inadequate with respect to Carnap's Bayesianism is suggested by the fact that talk of objectively determined probabilities entailed by the purported Bayesian-cum-extra construction of Carnap would virtually be pointless from the perspective of the purely subjective Bayesianism. However, my objective is not to analyze why the Bayesian-cum-extra view is wrong; rather the less demanding aim is to point out that since all the most devastating criticisms of Carnap's Bayesianism presuppose the Bayesian-cum-extra view, it might be worthwhile to explore an alternative reading of Carnap and see whether the rationale for his position is more fully appreciated.

There are two empiricist strategies that seem to seriously undermine Carnap's program of inductive logic. The first one attacks Carnap for neglecting empirical information in setting up the prior distribution. Objective and subjective Bayesians do not hesitate to join in contesting the "analytical" or "logical" status of Carnap's inductive probabilities. Remarkably, Creath emphasizes that these terms have particular meaning in Carnap's theory of knowledge, which empiricists characteristically fail to appreciate (Creath 1990, 6-7; 1991).[5] From this perspective the Bayesian identification of the latter would be altogether astonishing.

The second strategy mentioned in the foregoing undermines the thesis that the standard axioms of probability theory and the extra axioms are integrated together in Carnap's construction. It does so either on epistemological or on formal grounds. The epistemological point is to retain the special status of the standard axioms of probability theory, and challenge the justification of the extra axioms. The formal point is to represent all the axioms in a neat way so that their apparent epistemological countenance dissipates.

In order to give some substance to the two strategies I shall discuss them in more detail. Salmon develops several arguments against Carnap, most of which, however, can be subsumed under the first strategy (Salmon 1967a, 726-37; 1993).[6] He acknowledges the fact that the principal difficulty with Carnap's construction, as he understands it, is Hume's problem: two different events, for instance the observed and unobserved, are logically independent. Thus, Salmon is led to challenge the analytical character of Carnap's systems (Salmon 1967a, 732-33). For if the confirmation function is analytical, how, together with the synthetic evidence statement *e*, could it state anything about the hypothesis *h*? If there is no logical dependence between different events, then an analytical confirmation function cannot yield any prediction whatsoever. Salmon proceeds then to point out that since there is no logical relation between *e* and *h*, the resulting confirmation function, which logically relates *e* and *h*, must be a consequence of taking a particular decision as to a definition of the degree of confirmation. This decision is arbitrary, unless justified. But the forthcoming justification will have to be synthetic a priori.[7]

A somewhat similar point is addressed by Howson (Howson and Urbach 1993, 51-72). He claims that Carnap's definition of the degree of confirmation is undermined by Hume's problem, and presupposes what it is to prove. For the confirmation function will efficiently yield numerical values only if the parameter λ is specified. The parameter, the objection continues, cannot be univocally determined on logical grounds only; its value has to be set up by a decision. The behavior of the characteristic function, however, clearly suggests that λ measures how much one is willing to rely on empirical data (i.e. how much the confirmation of *h* on *e* approximates the relative frequency). The parameter then overtakes what the confirmation function was supposed to do, to wit, to measure the inductive support of *h* on *e*. Carnap's systems cannot be logical, for ultimately they depend on a decision assigning an arbitrary numerical value to λ, while logic remains "essentially noncommittal on substantive matters" (Howson and Urbach 1993, 71-72). In consequence, Howson also comes to the conclusion that Carnap's systems are not analytic. Why then is Carnap's confirmation function "committal on substantive matters"? Perhaps, because, as Howson claims, the definition results in priors, which exclude certain kinds of worlds, e.g. a completely chaotic world. Thus the elimination of a world is based on an arbitrary decision, which is arrived at despite ignorance of the relevant empirical information. At this point Howson makes a greater effort than Salmon to repudiate Carnap's purported arbitrariness. Even though he is ready to recognize that decisions form a natural part of the fabric of science, he nonetheless points out the important difference between the role decisions play in empirical science and the role they play in inductive logic. In science, he claims, decisions are "verified" by the adequacy of empirical representation. This could not be the case in inductive logic, which itself is supposed to "adjudicate decisions of empirical adequacy" (Howson and Urbach 1993, 70). Some authors tried to construe meta-inductive arguments to justify Carnap's prior distribution. Any argument, however, that acknowledges an empirical determination of the value of the parameter λ must, as Howson correctly claims, lead to inconsistencies in Carnap's system.

An important consequence of Howson's critique of Carnap is the conclusion that Carnap took probabilities to be logical quantities determined objectively, while in Howson's account they are to be understood as a result of subjective assessment. Thus, Carnap is located on the objective part of the objective-subjective Bayesian dichotomy. In the last section of this chapter, however, I find this conclusion wanting.

Let me now turn to the epistemological part of the second strategy. Salmon acknowledges that Carnap's rules generating the counterpart of the standard axioms of probability theory enjoy an elaborate pragmatic justification, for a subject which does not attain them will display clearly irrational behavior in bets.[8] For the other rules, however, Carnap would have to offer a non-pragmatic justification. The rules of invariance (e.g. the symmetry principle), for instance, are purportedly justified by Carnap on the ground that they represent a valid part of the principle of indifference, and ultimately in view of the fact that they reflect our inductive intuition. Or, the axiom of convergence is seemingly justified by the appeal to the fact that it is intuitively desirable that the confirmation function be capable of learning from experience. Salmon's general point, then, is that Carnap's attempts to justify the extra axioms invoke inductive intuition,[9] which he is not able to account for. I try to come to grips with this point in section 4 below.

Both Salmon and Howson, regardless of the foregoing criticisms of the extra axioms, do not embrace Fisher's skepticism with regard to the standard axioms of probability theory. Howson, for instance, demonstrates that the latter form necessary and sufficient conditions of consistency of conditional beliefs (Howson and Urbach 1993, 78-86).

Thus the first and second strategies of Salmon and Howson presuppose the Bayesian-cum-extra view on Carnap, which accounts for the delineation between the epistemic status of the standard axioms of probability theory and the extra axioms. Hence, in view of Carnap's inductive probabilities yielding non-satisfactory results, it is possible now to blame not the whole Bayesian-cum-extra set-up, but only its part consisting of the extra axioms, retaining thus the standard axioms of probability theory.

The criticisms discussed thus far purport to demonstrate the ultimate failure of Carnap's program. However, there are also radical objections raised by those who continue to work on Carnap's systems. Kuipers, for instance, emphasizes that from the perspective of the standard probability theory, Carnap's systems form a mixture of a purely formal framework and an application. Adopting the Bayesian-cum-extra view enabled Kuipers to accomplish the representation of Carnap's systems within the standard set-theoretical framework.[10] Seemingly, Carnap's initial formal constructions could thus be subsumed by a more powerful formulation, rich enough to study and compare a much wider class of inductive systems beyond the ones studied by Carnap.[11]

However, the piecemeal approach endorsed by the Bayesian-cum-extra view is insensitive to epistemological principles underlying Carnap's construction. Indeed, once we adopt the Bayesian-cum-extra view the latter are no longer under our grip (Skyrms 1991, 456). Initially Carnap constructed his systems in first order logic as semantical systems, based on the concepts of *state-description* and *range*; probability calculus was intended to form only a technical and philosophically insubstantial part of inductive logic thus understood, which was expected to be foundational with regard to statistical inferences. Carnap also advocated a dichotomy between two concepts of probability, i.e. the empirical, frequency concept of probability, and the logical concept of probability. In a formal framework devoid of semantical concepts (e.g. model, and truth in a model) this dichotomy apparently diminishes. In consequence, while the degree of confirmation is deprived of its meta-level and normative status, inductive logic is swiftly absorbed into probability theory and statistics, and particular systems are no different than specific statistical distributions (Kawalec 2000a). W. Sellars, for one, emphasizes the philosophical significance of the meta-level character of sentences "$c(h \mid e) = r$" stating the degree of confirmation of h given e (Sellars 1964, 220). Carnap—being aware of these consequences—was careful not to conflate set-theoretical with semantical representation, as he thought that set theory was not adequate as a complete logico-mathematical setting for a general language of science (Bohnert, 1975a, 210).

The Bayesian-cum-extra view, due to its epistemological inadequacy with respect to Carnap's *Logical Foundations of Probability* and *The Continuum of Inductive Methods* project, in one way or another undermines the latter. This fact motivates the search for grounds to defend the plausibility of the structural reliabilist interpretation of Carnap's Bayesianism I propound in this book. The sections to follow elaborate such grounds: I discuss how the structural reliabilist theory of justification is able to come to grips with difficulties of mentalistic reliabilism, I argue that on the structural reliabilist interpretation the systems of inductive logic are able to meet the Goodman paradox, and finally I examine the contribution of the structural reliabilist theory of justification to the debate between objective and subjective Bayesians.

2. Structural Reliabilism and Difficulties of Mentalistic Reliabilism

Some of the most important objections against mentalistic reliabilism were already discussed in the fourth section of the previous chapter. These are the problems concerning the very definition of reliability. In short, they refer to the extreme difficulty in picking out the relevant class of processes whose reliability is to be assessed. A further complication is due to the requirement that this specification of the relevant class of processes has to be accomplished in non-epistemic terms (Goldman 1979/1993).

Before I proceed to discuss further objections which are characteristic for the mentalistic reliabilism, I have to discuss some other difficulties which point out, in effect, that structural reliabilism could not a viable *epistemological* position at all. In other words, structural reliabilism has to face first these arguments, which reject all minimalist theories as *epistemological* standpoints. Sometimes the rejection concerns the question of the procedure of transforming a minimalist theory of justification into a mentalistic one (Chisholm 1982, 33-39). Only if these objections are successfully met can I proceed to discuss whether and how structural reliabilism is able to cope with objections characteristically directed against all sorts of reliabilism.

We may now ask what is the rationale for the radical neglect of all sorts of minimalist theories of justification? In general, it is widely assumed that the central task of epistemology is a theory and definition of knowledge. Remarkably, the standard definition of knowledge implies the entailment principle: knowledge entails belief. This principle seemingly rules out (MITJ), and undermines thus minimalist theories of justification. Beyond the standard definition of knowledge, however, others are possible too:

> I am not talking about knowledge in the abstract, as we sometimes do. We talk, for instance about the extent of "human knowledge." Some of this knowledge is solely in books, and not remembered by anyone. Thus, some scientific knowledge might be of propositions no one actually believes, propositions available to us should we need them, but not objects of actual belief. We can talk about the structure of such knowledge in the abstract, say about whether all the propositions of scientific knowledge can be systematized by certain basic laws of physics or chemistry, so that

these basic laws are like the axioms of geometry and the other laws, like its theorems, are derivable from the basic laws (Audi 1998, 179).

The tripartite definition of knowledge, which has a long and well-established tradition in epistemology (Chisholm 1982, 43); (Bronk 1998a, 71-72), involves at least three factors: truth, justification and belief. And since minimalist theories of justification do not embrace propositional attitudes they cannot, it is claimed, serve as a basis to the sketched definition of knowledge. Chisholm, for instance, characteristically admits that neither deductive nor inductive logic can confer evidence on its conclusions. Therefore, as such, both are irrelevant to epistemology (Chisholm 1966/1977, 64-65). Minimalist theories of justification dispense with beliefs in a way consonant with (MITJ) (see definition 2.6 on page 67), and therefore those theories are irrelevant from the point of view of the central task of epistemology, namely the task of providing the theory of knowledge, for on the standard view, as already noted, knowledge entails belief (Lehrer 1974, 12; Sosa 1991, 165; Hetherington 1996, 13-14).

One conceivable response could be that the task of providing a theory of justification is not intimately bound up with the task of providing a theory of knowledge, and that the two tasks can be kept separate. Admittedly, both Carnap and Hintikka seem to be far from endorsing this kind of strategy, and I shall not elaborate it here.

I will begin my defense with an attempt to substantiate the view that transmission of justification by non-epistemic structure is a genuine option and perhaps antecedent to mentalistic theory of justification. Only then I will examine the possibility of accommodating the minimalist theory of justification to theory of knowledge.

The starting point for Carnap is that the epistemic status of the logical (or mathematical) structure of the set of scientific sentences is unquestionable. Hempel offers the following explanation: "It is this purely deductive character of mathematical proof which forms the basis of mathematical certainty" (Hempel 1945/1949a, 240). Certainty then is conferred upon what is proved provided that is implied by the implicit definitions. In other words, the certainty results from the interplay between transition mechanisms and conventions. It warrants not only that the epistemic relations between elements of the structure can be well and precisely defined, but also that these are objectively constituted, and thus intersubjectively accessible. This is in virtue of the unquestionable epistemic status of the logical structure as well as the capability of the latter to induce uniquely, and indeed *quantitatively*, the epistemic status upon its constituent parts that sentences—atomic as well as complex—acquire their own epistemic status.

At the time that Carnap came up with the first monumental construction of the system of inductive logic a well defined semantical theory of the concept of the logical consequence was known. But this theory was only adequate for the specific area of justification-transmitting inferential relations among sentences. What was still needed was a broader, indeed a comprehensive, theory of such justification-transmitting inferential relations. Such a theory was still to be developed for non-demonstrative inferences, which constituted a dominant portion of relations among scientific sentences. And it seems that both Carnap and Hintikka agree that the construction of such a comprehensive theory is to precede a more complex theory of relations between propositional attitudes.

Carnap offers the following as an example illustrating the priority of a minimalist theory of justification

> Suppose we ask an engineer who is building a bridge why he has chosen the building materials he is using, the arrangement and dimensions of the supports, etc. He will refer to certain physical laws ... On further inquiry as to his confidence in these laws he may apply to them phrases like "very reliable", "well founded", "amply confirmed by numerous experiences". What do these phrases mean? It is clear that they are intended to say something about ... degree of confirmation (Carnap 1945/1970, 467-68).

Therefore, a comprehensive theory of epistemic status of *all sorts of sentences* of a given language, in particular the language of unified science, must be developed first, and only then can we proceed to the task of constructing a theory of *doxastic propositional attitudes* (Carnap 1938/1949, 408-10; 1946a; 1955/1956, 250).

J. Fodor claims that Carnap's account of propositional attitudes as relations between people and sentences in (Carnap 1946a) is satisfactory both on *a priori* and empirical grounds (Fodor 1978/1981, 51-56). Nevertheless, since (Carnap 1962) onwards Carnap advanced the allegedly opposite view that a minimalist theory of justification is to be developed by an idealization of the conditions stated on the grounds of a mentalistic theory, and that the distinction between these two kinds of theories can be blurred.

Although not explicitly advanced by Carnap, an argument for the priority of the construction of a minimalist theory over the construction of a mentalistic theory is the following:[12] in the construction we have to draw upon the results of the relevant sciences. In this particular case the relevant science is *the science of logic*, which was developed, at that time, mainly by A. Tarski, K. Gödel and Carnap himself. However, it was a well developed theory for *semantical* notions only, but not for *pragmatic* ones. And Carnap was strongly convinced that there was *no logical theory* for pragmatic notions. He thought that pragmatics was *an empirical science* (Carnap 1955/1956, 250; Martin 1959, 3; Koj 1987, 282).

N. Cocchiarella provides an excellent historical explanation why Carnap assumed pragmatics to be an empirical science:

> Prior to the 1960s most formal language philosophers deliberately avoided the whole question of how the different contexts in which language is used affects the meaning and reference of expressions. Logicians, with their proper concern for rigour and formal precision, tended to see the logical analysis of those features of ordinary language that were dependent on context as either impossible or hopelessly complex. They preferred instead to abstract from particular contexts of use and to provide analyses of sentences and arguments of ordinary language within a timeless and context-free setting.
>
> Logical semiotics at that time, as opposed to descriptive semiotics, was in fact by definition restricted to logical (or pure) syntax (concatenation theory) and logical (or pure) semantics (model-theory as a branch of set-theory or as a form of higher-order logic), where the latter amounted in effect only to a theory of reference. Pragmatics, the third branch of the semiotical trinity, was not only excluded from logical or pure semiotics on this scheme, but was in fact considered to be the very basis of descriptive semiotics (Cf. Carnap, R. *Introduction to Semantics*. Cambridge: Harvard University Press, 1942). Thus Rudolf Carnap, for example, considered pragmatics to be an empirical theory which was to be formulated in the metalanguage of a collection of object languages and which was in that regard to be concerned with the relationship between the users and the signs of the object languages in question. The expressions of a pragmatic theory, on this view, including epistemic expressions for belief, doubt, etc., were considered to be *predicates* belonging to the metalanguage and were understood as such as designating psychological relations between language users and signs of the object languages. Indeed, Carnap clearly states even as late as 1963 that the relation of designation in pragmatics "is a psychological concept" (cf. Carnap, R., *Replies*. In P. A. Schilpp, Ed., *The Philosophy of Rudolf Carnap*, p. 125. La Salle: Open Court, 1963) (Cocchiarella 1981, 116-117).

From Carnap's epistemological standpoint the difference between the empirical nature of pragmatic concepts and the logical nature of semantical concepts was crucial. Logic warrants the epistemic status only of its constituent parts, and the results of empirical science acquire epistemic status insofar as the logical structure is imposed upon them. Hence, the results of empirical investigation can in no way contribute to solving *epistemological* questions. Therefore, the actual behavior of empirical cognitive agents is irrelevant from the epistemological perspective. Thus, in effect, we can proceed to develop a mentalistic theory, but this kind of theory will never be *an epistemological theory* unless we already have a minimalist theory, upon which the mentalistic theory will rest.

There are, however, many objections to the effect that the theory of justification that Carnap and Hintikka project in their systems of inductive logic could never serve in building a theory of knowledge. The former theory—for many reasons—is inappropriate for the main task of epistemology, namely for constructing the theory of knowledge. In order to provide the conditions for knowledge we need an already well-worked out theory of truth, of justification and of belief. Various objections seem to suggest that the theory of justification advanced by Carnap and Hintikka is either reducible to the theory of truth, or is in conflict with the concept of belief. I will proceed now to discuss these two kinds of objections in turn.

In the first chapter I explained in detail what precisely makes the Carnap-Hintikka style systems of inductive logic *semantical*; roughly, the systems are intended to base the condition of justification on some elementary concepts—such as *state-description* (model), *range* or *constituent*—which are, in turn, defined as semantical concepts.[13] Does this not entail that the theory of justification proposed by Carnap and Hintikka reduces to the theory of truth? But if on Carnap's and Hintikka's account the definition of truth is sufficient to provide the condition of justification, then a theory of justification embracing such a condition is not pertinent to the theory of knowledge. In fact, Carnap readily admits those limitations in applying his inductive logic to knowledge, which are due to the simplicity of language of the systems of inductive logic as well as a very simple evidence statements expressible therein (Carnap 1947/1997a, 1-2). In sum, the condition of justification required by the theory of knowledge must then contain an ingredient which cannot be stated in terms of truth, or the related concepts. Carnap, for instance, delineates the difference between the concept of justification and truth in a way consonant with the standards of contemporary epistemology; in his terms, the difference is analogous to the difference between the fact that p is true and the fact that criteria of evaluation of whether p is true or not are epistemically available for S (Carnap 1936/1949, 120; 1946; Ingarden 1962).

What might this extra "ingredient" of the definition of justification be? Presumably, the mainstream epistemologists will suggest that the concept of justification must relate to *evidence*. Epistemology is the theory of knowledge of limited empirical agents. And their knowledge depends on the evidence that is in possession of these agents.[14] Their acts of cognition are always involved in a particular context, assuming social, historical and cultural perspective, and their merit does not surpass the merit of a cognitive apparatus that is in the disposal of a cognitive agent. The attribution of "knowledge" in our common linguistic practices is, in one way or another, sensitive to these limitations. Furthermore, it is this common linguistic practice that cannot be ignored in the theory of knowledge that we are to construct. Thus, the concept of evidence must be accounted for in the proposed concept of knowledge.[15] And since neither the concept of truth nor of belief is appropriate, the only way to account for the concept of evidence is the theory of justification. In Carnap's and Hintikka's terms it means that it is not possible to contrive a concept of justification which will be suitable for the concept of knowledge we seek to define in epistemology, unless it is a *pragmatic* concept.[16] The epistemic subject of knowledge—and by the same token the epistemic subject of justification—must be referred to in the application condition of justification formulated by a given theory of justification. As Carnap, however, assumes that pragmatic concepts are empirical, and cannot serve to define epistemological concepts, it would seem that there is no way he could appropriate his semantical, and by the same token minimalist, theory of justification to make it a pragmatic one. Therefore, the minimalist theory of justification proposed by Carnap and Hintikka in their systems of inductive logic would be irrelevant for epistemology, for this theory could neither be one of the building-blocks of the theory of knowledge, nor could it be appropriately modified for this purpose.

There are many ways for Carnap and Hintikka to get around this difficulty. They might, for instance, claim that they are not interested in constructing any theory of knowledge, or that the theory of knowledge they project is very different from the widespread theory of knowledge, which requires recourse to evidence and empirical cognitive agents. This, in fact, will not be a satisfactory answer, and it would even be hard to actually find support for such claims in the writings of Carnap and the members of the Finnish School. I suggest that the discussed difficulty can be met on its own ground as follows.

What does, in effect, the objection that the condition of justification proposed by Carnap and Hintikka is definable in terms of truth mean? Roughly, on their account any sentence is expressible as a disjunction of the descriptions of those states of affairs, in which a sentence obtains. Since we know precisely what are *all* possible states of affairs, we can calculate the ratio of the number of states of affairs, in which a given sentence obtains, to all possible states of affairs. If the ratio is 1, then the sentence is *logically true*; if it is 0 then it is *logically false*, to wit, it is a tautology or, respectively, a counter-tautology. Similarly, we can calculate the ratio of the number of states of affairs, in which the conjunction of sentences A and B obtains, to the number of states of affairs, in which the sentence B obtains. If the number is 1, then B logically implies A; if it is 0 then B logically implies $\neg A$. There are, however, sentences which are neither logically true nor false—Carnap calls them *factual* sentences. And there are also types of relations between sentences, which are not conveyed by the two kinds of relations depicted above.

Carnap's and Hintikka's ambition was to provide a condition of justification for all sorts of sentences and their mutual relations in a given language. In order to achieve this, however, they had to go beyond merely semantical concepts. For instance, a metric convention was adopted, which led to uniform distribution of numerical values to all structure-descriptions (Carnap) or constituents (Hintikka). The convention also controls the way in which the up-coming evidence affects the existing distribution. Allegedly, these, and other, extra-logical, assumptions are not derivable from a definition of truth, and thus the argument fails. The condition of justification proposed by Carnap and Hintikka is not derivative of the concept of truth.

Moreover, the extra-logical—and by the same token the extra-semantical—assumptions do not uniquely determine the exact behavior of the confirmation function in the Carnap-Hintikka style systems of inductive logic. Thus, the way to construct a mentalistic theory of justification on the basis of this minimalist theory is open. Carnap is prone to think of a mentalistic theory of justification as arising in the context of application of the minimalist theory of justification. The application of the proposed theory of justification, and (RSR) in particular (see definition 2.10 on page 76), requires then the specification of *total evidence* available at the time of application as well as the specification of the domain of application (Carnap 1947/1997a, 6-9); (Swinburne 1997, 160-67). Even the very definition of the confirmation function is viable to the context-sensitive determination, for instance it is the task of the researcher to provide exact numerical values to the parameters of the systems (e.g. α or κ). Thus, the argument that in Carnap's and Hintikka's account the theory of justification is reducible to the theory of truth, and hence inappropriate for developing the theory of knowledge, is ungrounded.

We turn now to the second sort of argument to the effect that the minimalist condition of justification (RSR) cannot be made compatible with the mentalistic condition of justification which refers to the concept of belief. For there might be well defined justificatory relations—in the sense projected by Carnap and Hintikka—between sentences, which no one is able to apprehend or willing to believe. On the other hand, there might be sentences that most people believe, and think they do so justifiably, while it is not possible to assess the justificatory status of these sentences within Carnap's and Hintikka's construal.

Consider, for instance, two sentences "The one million and first swan is white" and "The one million and second swan is white". These are very simple sentences, and they are well-formed sentences in a very simple inductive system. Moreover, the system allows us to provide the numerical value of the justificatory status of one sentence conditional on the other, and numerical values of the justificatory status both sentences enjoy prior to any observation. However, it is psychologically implausible that someone might: (i) believe this sort of sentence, which keeps track of the exact number of the kinds of objects observed thus far; (ii) assign to sentences which one believes precise numerical values representing the degree of belief; (iii) assign precise numerical values to the relations between sentences. Therefore Carnap's and Hintikka's theory of justification seems to be in conflict with the concept of belief. Moreover, the conditions of knowledge the theory of knowledge is to provide have to accord with our intuitions—as required by *quietistic* character of epistemological theories—concerning the concept of knowledge. Thus, the condition of justification contrived in the systems of inductive logic would be dispensable for it seems to conflict with our intuitions concerning the concept of belief—the other ingredient of the definition of knowledge.

There are also sentences, so the argument continues, which we normally accept but which cannot be subsumed under the condition of justification (RSR) provided on the grounds of the systems of inductive logic. For instance, "John watches TV every evening", "There is a sheep in the field" or "I believe what you say is true". These simple examples illustrate that an important portion of our ordinary beliefs is not accounted for by the theory of justification projected by Carnap and Hintikka. Perceptual beliefs, e.g. "There is a sheep in the field" form a very elementary class of beliefs, as they condition the acquisition of non-perceptual kinds of beliefs such as, for instance, inferential beliefs. Higher order beliefs and relational beliefs also have to be accounted for by a theory of justification, which will then be employed to form a theory of knowledge.

The simplistic answer to the two objections discussed above is that neither Carnap nor Hintikka try to make their theory of justification compatible with the concept of belief. From the outset they project the minimalist theory of justification rather than the mentalistic theory. Such a straightforward answer will not do, however, for the following reason. Even if Carnap and Hintikka endorse a minimalist theory of justification they both expect that it can be further extended to a mentalistic theory of justification. And it is a particular context of application of the minimalist theory of justification to fill in the missing details to generate a mentalistic theory of justification. So, Carnap and Hintikka, in principle at least, project also a mentalistic theory of justification. Therefore, the formerly discussed objections still need to be answered.

Carnap's response to these objections turns on two basic arguments, both of which were to some extent addressed in section 4 of the previous chapter. Let us begin with the argument from the history of geometry. The moral from the history of geometry is that we need to embrace idealizations in order to come up with interesting results. Even if the projected minimalist theory of justification would be insufficient to match precisely all our intuitions concerning beliefs, it still might lead to important results. For the intended field of application of this theory of justification is science, and not our ordinary beliefs. How Carnap can proceed from an account of justification in science to an account of justification suitable for our ordinary beliefs is made perspicuous by the second argument. The argument from filling a circle with squares admits that *one* method might not be sufficient to cover all points at the same time, but for every point within the circle there is a method leading to covering the point with a square. Similarly, it might not be quite the same minimalist theory of justification that Carnap projects to be useful in science that will also be capable of covering all our intuitions concerning our ordinary beliefs, but still for all important domains of discourse there might be a minimalist theory of justification, which will, nonetheless, be suitable. For instance, appropriate modifications in the condition of quantitative justification (Q) (definition 2.12), or in the condition of minimalist justification (MITJ) (definition 2.6) may afford us such a minimalist theory of justification.

The argument from the history of geometry addresses the first kind of objection concerning the psychological implausibility of the assumptions endorsed by Carnap's and Hintikka's theory of justification. For if we are to construct a model of justificatory relations among sentences or beliefs there must be some idealizations. And the kind of idealization endorsed by Carnap is important insofar as it imposes a metric convention, which—as argued in section 4 of the previous chapter—also imposes objectivity upon the data.

The argument from the analogy of filling a circle with squares concerns the second sort of objection, to wit, that there are many important kinds of ordinary beliefs, which the minimalist theory of justification proposed by Carnap and Hintikka does not account for. However, both Carnap and Hintikka overtly admit that such inductive systems, which will be able to account for relational or higher order sentences, or beliefs, are still to be constructed.[17]

Neither of the two arguments, however, addresses the issue of justification of non-inferential beliefs, e.g. of perceptual beliefs. At this point I must admit that neither Carnap nor Hintikka undertake a discussion of this issue against the background of inductive logic. It is not an objection, however, if we assume—as I do throughout this book—that they propose the theory of *inferential* justification.

3. Structural Reliabilism and the Paradoxes of Confirmation

The paradoxes of confirmation may well serve as a background against which to test the epistemological limitations of a theory of confirmation. Out of the large number of paradoxes I will discuss those pertinent to the present argument.[18] I will argue that the structural reliabilist theory of justification endorsed in the systems of inductive logic is not only capable of resolving most important paradoxes of confirmation, but also that, at least in part, it is motivated by the challenge posed by the paradoxes. Thus, the solution of the paradox of statistical syllogism, for instance, motivates the prevailing concept of total evidence, which is indispensable to project a way for developing a mentalistic theory of justification and a theory of knowledge which would draw upon the minimalist theory of justification embraced by Carnap and the Finnish School. The solution of the ravens paradox, on the other hand, provides an additional motivation to develop a *quantitative* theory of justification rather than a qualitative or comparative one.

Most importantly, however, I will argue that the structural reliabilist theory of justification is immune to Goodman's grue paradox. This paradox was brought about with the intention of undermining the whole project I have been discussing so far, to wit, the project of developing a theory of justification on purely formal grounds. In effect, Goodman intended his paradox to demonstrate that the theory of justification embraced in the systems of inductive logic—which I argued to be the structural reliabilist theory of justification—is useless for a theory of knowledge, or epistemology in general.[19] And the reason is, roughly, that it is impossible—as the grue paradox is intended to demonstrate—to develop a non-pragmatic theory of justification. Lehrer, for instance, concludes the Goodman paradox thus: "[w]e cannot base justificatory inference on the form of the inference alone" (Lehrer 1974, 133).[20] I claim that the structural reliabilist theory of justification developed in the Carnap-Hintikka style systems of inductive logic is able to meet the Goodman paradox. By the same token I demonstrate that the structural reliabilist theory of justification took off as one of the viable alternatives, in fact an unprecedented one, among the contemporary attempts to provide the theory of justification.

Let me start with the paradox of statistical syllogism.[21] Consider the following form of an argument:

a is F (DS)

$\mathrm{P}(G \mid F) = 1$

Hence, certainly (or, necessarily) a is G

This is a form of a deductive argument, in fact, a syllogistic form. If the premises are true, then there is no alternative argument of the same form, in particular one with a different set of true premises, which will lead to a conclusion inconsistent with "a is G". Analogously, we can formulate an argument of the following form:[22]

a is F (SS)

$\mathrm{P}(G \mid F) \approx 1$

Hence, almost certainly (or probably) a is G

Call any argument of the form (SS) *statistical syllogism.* Statistical syllogisms, in opposition to arguments of the form (DS), lead, however, to difficulties. Let us consider the following arguments:

John is in this room. (SS1)

90 persons out of 100 present in this room are democrats.

Hence, almost certainly John is a democrat.

and

John is sitting in the last row. (SS2)

9 persons out of 10 sitting in the last row are not democrats.

Hence, almost certainly John is not a democrat.

In opposition to (DS), there are arguments of the form (SS), which have true premises, but which lead to conclusions inconsistent with the premises. This is the so-called *paradox of statistical syllogism.*

The paradox, Hempel claims, is based on the misleading use of the phrase "certainly". The application of this phrase in the conclusion alone obscures the logic of the inferences considered. For the way this phrase is applied suggests that it is a modal qualification of the conclusion alone, while, in fact, it is indicative of the relation that obtains between the premises and the conclusion. An important complement to Hempel's argument is the discussion that the usage of "certainly" as a putatively modal qualification of the conclusion of the statistical syllogism does not match any ordinary usage of this phrase (Hempel 1962, 134-36). The most appropriate explication of the phrase "certainly" in this particular usage, i.e. as indicative of the relation between premises and conclusion of a statistical syllogism, is the concept of logical probability "$c(.\,|\,.)$" developed in the systems of inductive logic. Thus appropriated form of the statistical syllogism is the following:

$$c(a \text{ is } G \mid \{a \text{ is } F \text{ and } \mathrm{P}(G \mid F) \approx 1\}) \approx 1 \qquad \text{(SSC)}$$

The conclusion of this form of statistical syllogism is analytical, and cannot lead to inconsistencies, similarly as the deductive argument of the form (DS) does not, provided the premises are true.

There is an open problem, however. If two sets of premises in deductive arguments of the form (DS) lead to two conclusions inconsistent with each other then we may infer that at least one of the sets of premises contains a false sentence. This is not, however, the case with arguments of the form (SSC). And in order to solve this problem Carnap gives birth to the concept of *total evidence*. Total evidence is the conjunction of all sentences relevant to the given context of application,[23] e.g. the context of an argument. If the total evidence is consistent, then the probability of any two sentences contained within the total evidence add up to 1. Therefore, the concept of total evidence is successful in eliminating the possibility of there being two inconsistent sentences, each of which could have a high degree of probability.

A. Ayer advances a criticism against this concept of total evidence, however. He claims that the degree of confirmation of a given hypothesis will vary respective to what portion of the relevant evidence is brought about. Moreover, any sentence of the form "$c(h \mid e) = r$" is an analytic truth in the systems of inductive logic. Hence, all the sentences "$c(h \mid e_1) = r_1$" throughout "$c(h \mid e_i) = r_i$" are necessary truths, even though $r_1 \neq \ldots \neq r_i$. Ayer is willing to grant that the total evidence e encompasses all the relevant pieces of evidence brought about by $e_1, \ldots, e_i$. Still, he claims, the sentence "$c(h \mid e) = r$" is the same kind of analytical sentence as any of the sentences "$c(h \mid e_1) = r_1$",..., "$c(h \mid e_i) = r_i$". In sum, it is not possible to justify the requirement of total evidence in inductive logic alone (Ayer 1957, 12-17). The point, however, is granted by Carnap himself, who admits that the requirement of total evidence is not a logical requirement, but a *methodological* requirement, i.e. the requirement of total evidence is a necessary condition for an *application* of inductive logic in a given context (Carnap 1950, 202-208). Admittedly, there are serious problems we encounter when trying to apply this requirement (Chisholm 1982, 37). But still, as Hempel claims, there are many examples of explanatory and predictive arguments in empirical science, which satisfy the requirement of total evidence (Hempel, 1961, sec. 11).

Another paradox to be considered is the ravens paradox which was formulated first by Hempel in (Hempel 1945/1965, 14-20).[24] On the one hand, it seems natural to require that two hypotheses, which are logically equivalent, be confirmed by the same evidence. For instance, the hypothesis H that "All ravens are black" has to be confirmed by the same evidence that the hypothesis "All non-black things are non-ravens". On the other hand, however, it seems counterintuitive—as regards our intuitions on "confirmation"—to admit that an observation of white shoes confirms the hypothesis that "All ravens are black".

There have been many attempts to resolve the paradox. The first one to make itself reputation was the solution proposed by J. Hosiasson-Lindenbaum (Hosiasson 1940).[25] Her idea was to introduce a *quantitative* concept of confirmation.[26] Thus, an observation of a non-black non-raven does, indeed, confirm the hypothesis "All raven are black", but to a much smaller extent than does an observation of a black raven. As it happens, the number of ravens is much smaller than the number of non-black things. Granted this fact, she proves then that:

$$c(\text{“}x\text{ is a non-raven”} \mid \text{“}x\text{ is a non-black thing”}) > c(\text{“}x\text{ is a black thing”} \mid \text{“}x\text{ is a raven”}).$$

In consequence, the observation of a black raven raises the confirmation of the hypothesis "All ravens are black" to a greater extent than does an observation of a non-black non-raven (Hosiasson 1940, 137-38).[27]

There are, however, some flaws in the proof that Hosiasson constructs.[28] For the sake of simplicity, in presenting, and criticizing her argument, let me use the standard probability function instead of the confirmation function, which is then defined by her as conditional probability.[29] In the argument Hosiasson refers to the theorem:

> If the evidence E is a consequence of the hypothesis H and $0 < P(H) < 1$, then the value of $P(H \mid E)$ increases as the value of $P(E)$ decreases (Hosiasson 1940, 133; Mortimer 1982, 187).

Neither of the two evidence sentences "x is a raven and x is a black thing" and "x is a non-raven and x is a non-black thing", however, is a consequence of the hypothesis "All ravens are black". For the argument Hosiasson advances it is necessary to state then what is the probabilistic dependence of the two evidence statements upon the hypothesis, because—according to Bayes's theorem (see page 5)—the value of $P(H \mid E)$ depends, if $P(H)$ is given, not on $P(E)$ alone, but on the ratio $P(E \mid H) / P(E)$.[30] But still, Hosiasson's argument can be recovered if we explicitly assume that sentences "x is a raven" and "x is a black thing" are probabilistically independent of H. The hypothesis "All ravens are black" is logically equivalent to the sentence "For all x, x is a raven and x is a black thing iff x is a raven". Hence—by equivalence (proposition 1.8 b)—for all x it is the case that $P($"x is a raven and a black thing"$) = P($"x is a raven"$)$. To demonstrate Hosiasson's point let me assume that:

$$0 < P(H) < 1, \tag{i}$$

and also that our prior knowledge includes the information that non-black things outnumber ravens:

$$P(\text{“}x\text{ is a raven”}) < P(\text{“}x\text{ is a non-black thing”}).$$

Again, by equivalence we obtain that:

$$P(\text{“}x\text{ is a black raven”}) = P(\text{“}x\text{ is a raven”}) - P(\text{“}x\text{ is a non-black raven”}).$$

Similarly,

$$P(\text{“}x\text{ is a non-black non-raven”}) =$$

$$P(\text{“}x\text{ is a non-black thing”}) - P(\text{“}x\text{ is a non-black raven”}).$$

From the three statements above it follows that

$$P(\text{“}x\text{ is a black raven”}) < P(\text{“}x\text{ is a non-black non-raven”}). \tag{ii}$$

The hypothesis "All ravens are black" is inconsistent with the sentence "x is a non-black raven". Hence,

$$P(\text{“}x\text{ is a black raven”} \mid \text{“All ravens are black”}) =$$

$$P(\text{“}x\text{ is a raven”} \mid \text{“All ravens are black”})$$

and

$$P(\text{“}x\text{ is a non-black non-raven”} \mid \text{“All ravens are black”}) =$$

$$P(\text{“x is a non-black thing”} \mid \text{“All ravens are black”}).$$

Given the probabilistic independence, we obtain that

$$P(\text{“}x\text{ is a black raven”} \mid \text{“All ravens are black”}) = P(\text{“}x\text{ is a raven”})$$

and

$$P(\text{“}x\text{ is a non-black non-raven”} \mid \text{“All ravens are black”}) =$$

$$P(\text{“}x\text{ is a non-black thing”}).$$

Now, we can compare the probabilities of the hypothesis “All ravens are black” conditional on the two different observations: “x is a black raven” and “x is a non-black non-raven”:

$$P(\text{“All ravens are black”} \mid \text{“}x\text{ is a black raven”}) =$$

$$[P(\text{“All ravens are black”}) \cdot P(\text{“}x\text{ is a raven”})] \mid P(\text{“}x\text{ is a black raven”}) =$$

$$P(\text{“All ravens are black”}) \cdot (p_1 + p_2) / p_1; \qquad \text{(A)}$$

where

$$p_1 = P(\text{“}x\text{ is a raven and }x\text{ is black”});$$

$$p_2 = P(\text{“}x\text{ is a raven and }x\text{ is a non-black thing”}).$$

Similarly,

$$P(\text{“All ravens are black”} \mid \text{“x is a non-black non-raven”}) =$$

$$[\mathrm{P}(\text{“All ravens are black”}) \cdot \mathrm{P}(\text{“x is a non-black thing”})] /$$

$$/ \mathrm{P}(\text{“x is a non-black non-raven”}) =$$

$$P(\text{“All ravens are black”}) \cdot (p_4 + p_2) / p_4; \qquad \text{(B)}$$

where

$p_4 = P($"x is a non-raven and x is a non-black thing").

The value of (A) is greater than the value of (B), only if

$p_1 < p_4$ and $p_2 > 0$.

If the latter would not be satisfied then it holds trivially that

P("All ravens are black") = 1.

The former is satisfied according to (ii).

Hosiasson's result it not directly applicable to the systems of inductive logic discussed in this book, because there is no straightforward way to incorporate the *factual* assumption (ii)—which is essential for this type of solution—into the systems of inductive logic embracing symmetry conditions, especially the symmetry conditions for predicates (definitions 1.8 and 1.11, propositions 1.21 to 1.26).[31] For the type of language we applied to discuss the paradox, the symmetry between primitive predicates would require that instead of the inequality in (ii) we have:

P("x is a black raven") = P("x is a non-black non-raven").

Suppes comments precisely on this point:

> Now the inductive logician may reply that if he were seriously pursuing this example, he would begin much further back and introduce a number of fundamental elementary predicates into his language, and use them to express explicitly the known evidence about ravens ...
>
> And this reply leads to the second Bayesian retort. The inductive logicians, to the Bayesian at least, are the heirs to an intellectually distinguished but misguided tradition of logical atomism that begins at least with Hume. ... [F]or Bayesians it is an approach that is bound to fail for fundamental reasons. First, it is impossible to express in explicit form all the evidence relevant to even our simplest beliefs. ... [Second,] [i]t is of fundamental importance to any deep appreciation of the Bayesian viewpoint to realize that the particular form of the prior distribution expressing beliefs held before the experiment is conducted is not a crucial matter. ... The well-designed experiment is one that will swamp divergent prior distributions with the clarity and sharpness of its results, and thereby render insignificant the diversity of prior opinion (Suppes 1966a, 202-204).[32]

Carnap does not provide an elaborated solution to the paradox. However, he admits that he would embrace Hempel's own solution (Carnap 1950, 469-70), and that the possibility of constructing a quantitative system of inductive logic turns on this kind of solution (Carnap 1950, 223-24). What is wrong then with Hosiasson's result as a motivation to construct the systems of inductive logic as *quantitative* rather than *qualitative* theories is essentially the fact that we cannot embrace the assumption (ii) on the grounds of inductive *logic* alone. The assumption (ii) derives from our knowledge of the world, and it is not a matter of logic to determine empirical facts. The quantitative measure function defined on the grounds of inductive logic cannot be based on empirical data such as this reported in (ii), or any other pragmatic presuppositions.[33]

Admittedly, the ravens paradox does not constitute a genuine paradox on the grounds of inductive logic alone. The paradox arises only when we bring in extra information. Someone might claim that "*x* is a non-black non-raven" does not constitute a confirming instance of the general hypothesis "All ravens are black". But suppose we device two different kinds of tests to check whether an object is a raven and whether it is black. Suppose further that we examine an *unspecified* kind of object, and, after performing the latter test, we acknowledge that this object is not black. Finally, after performing the former test, we acknowledge that this object is not a raven either. In this account it seems that indeed an instance of non-black non-raven constitutes a positive instance of the hypothesis "All ravens are black", provided we do not assume that the kind of object we put under the tests is known to us in advance to performing the test.

Suppose, however, that prior to performing the tests we know that the examined object *x* is a non-raven. Since from the hypothesis "All ravens are black" it follows that "*x* is non-raven or *x* is black", then the evidence we possess prior to performing the tests implies the latter consequence of the general hypothesis, and therefore we know that this instance is a confirming instance of the hypothesis *prior* to performing the tests. When we device the system of inductive logic, however, we cannot assume which empirical information would be available (Scheffler 1963/1967, 265-67).

Among the traditional epistemologically motivated objections to Carnap's program it is undoubtedly the Goodman paradox[34] that aims at undermining the relevance of Carnap's—and by the same token the relevance of the Finnish School's—construal to epistemology. In short, Goodman's idea was to demonstrate that it is impossible to construct *formally* a theory of justification, for any such theory has to be cast in pragmatic terms. Therefore Carnap's and Hintikka's systems of inductive logic would be inappropriate for epistemological purposes from the outset, as they are constructed by means of *syntactical* and *semantical* terms alone

(Carnap 1938/1949, 408-10). Carnap overtly acknowledges that: "inductive logic is, like deductive logic, in my conception a branch of semantics" (Carnap 1945/1970, 452), and that semantical concepts, in particular the concept of state-description, constitute a sufficient basis to avoid the Goodman paradox (Carnap 1947/1997b, 22).[35] Let me now discuss the Goodman paradox in some detail.

The initial version of the paradox, which Goodman presented in 1946 (Goodman 1946), runs as follows: suppose that we draw balls from an urn; one ball per day. Suppose that we have been drawing the balls for ninety nine days including Victory in Europe Day, May 8, 1945. And all the balls that have been drawn so far are red. The evidence we have supports the expectation that the next ball to be drawn from the urn will also be red. However, there is a predicate "*S*", which Goodman defines as "is drawn by Victory in Europe Day and is red, or is drawn later and is non-red". The predicate *S* applies to the ninety nine balls drawn so far—they have been drawn by Victory in Europe Day and all of them have been red. Hence, all ninety nine balls are *S*. But the predicate *S* leads to a new prediction, indeed, one inconsistent with the previous one, namely, that the next ball to be drawn will not be red. The expectation that the next ball will be non-red is counterintuitive, as we naturally expect the next ball to be red. But it turns out that evidence supports both expectations equally well. Let e_R stand for the evidence we have after drawing the ninety nine balls, which is formulated with the predicate *R* "is red" (similarly, e_S is the same evidence cast in terms of "*S*"). Let $c(.\,|\,.)$ be a confirmation function defined in one of the Carnap-Hintikka style systems of inductive logic. Thus, we have

$$c(R(a_{100})\,|\,e_R) = r_1,$$

$$c(S(a_{100})\,|\,e_S) = r_2.$$

The mathematical function $G(s_i, s, \kappa)$ that "completely" characterizes the confirmation function depends on the number of primitive predicates κ, the number of observed instances of a given predicate s_i, and the total number of individuals observed s. Now, it would seem that $\kappa_R = \kappa_S$, $s_R = s_S$, and s is obviously the same throughout both descriptions. From these equalities it follows, on any exclusively semantical definition of confirmation based on the characteristic function $G(s_i, s, \kappa)$, that $r_1 = r_2$.[36] We know, however, that

$$S(a_{100}) \rightarrow \neg R(a_{100}),$$

so that the two inconsistent predictions have the same degree of confirmation. This apparently demonstrates the inadequacy of any definition of confirmation that Carnap is able to come up with, because these definitions are incapable of discerning between the *S*-type and the *R*-type predicates.[37]

Carnap's own objection was, in effect, that "$\kappa_R = \kappa_S$" does not hold (Carnap 1947/1997a, 14-15).[38] The reason is that "*S*" is a quantitative predicate, i.e. one that requires time-reference, while "*R*" is a qualitative one. This objection, however, as Goodman correctly points out, will not do.[39] For "*R*" might be defined by means of "*S*" as "is drawn by Victory in Europe Day and is *S,* or is drawn later and is non-*S*".

It was in 1954 that Goodman's book, entitled *Fact, Fiction and Forecast*, appeared (Goodman 1954/1983, 72-81). And it was this book that contained the wide-spread version of the Goodman paradox. Suppose that all emeralds examined before *t* have been green. This evidence, thus, supports the hypothesis that all emeralds are green as well as the instance-prediction that the next emerald to be observed, i.e. to be observed after *t*, will also be green. Suppose, however, that we introduce a predicate "Grue" defined as "is examined before *t* and is green, or it is not examined before *t* and it is blue". By the evidence we have, all emeralds observed so far are grue. And the evidence supports the instance-prediction that the next emerald to be observed will also be grue. An analysis analogous to the one above will demonstrate that on any definition of the confirmation function provided by Carnap both predicates "Green" and "Grue" are indistinguishable. Thus, the formal definition of the confirmation function is indifferent to the paradox that arises. It will only be possible to discern between the two kinds of predicates, Goodman claims, in pragmatic terms. Thus, if we want to admit that only the "Green" hypothesis and "Green" prediction are confirmed, while the "Grue" hypothesis and the "Grue" prediction are not, we have to introduce a term of distinction. It cannot be a syntactical or semantical term, for the failure of Carnap's systems of inductive logic to cope with the distinction illustrates that these sorts of terms will not do. Goodman's own solution was to involve "entrenchment". What distinguishes "Green" from "Grue" is that the former is entrenched in our practices of prediction-making and hypothesis-confirmation, while the latter is not.

It is not my aim here to discuss the viability of Goodman's own solution of the paradox. My present concern is whether the Goodman paradox is resolvable on the grounds of the Carnap-Hintikka style systems of inductive logic.[40] The issue is crucial for the epistemological interpretation of the systems I am propounding throughout this book. If the Goodman paradox is not resolvable than the degree of confirmation cannot be a candidate explication for the condition of justification. For if the Goodman paradox holds in the systems of inductive logic, then the *same* degree of confirmation is assigned to two hypotheses, one of which, i.e. the "Grue" one, we consider as *unjustified*. In such a case, the degree of confirmation would be the same for justified and unjustified sentences. The point could be made even more strongly: it is not only Carnap's and Hintikka's attempts to provide the minimalist condition of justification that fail—none at all can be given. And the reason is that any minimalist condition of justification has to be formulated in syntactical and semantical terms alone, and any such attempt would be susceptible to the Goodman paradox, precisely for the reason that it is impossible to avoid the paradox without invoking pragmatic terms.[41] The more general point, to wit, the skepticism towards any minimalist condition of justification, has been welcomed by most contemporary epistemologists (Lehrer 1974; Sosa 1991).

The argument I am going to present demonstrates that the Goodman paradox does not arise in the Carnap-Hintikka style systems, and therefore, the epistemological relevance of the systems is not threatened thereby. A defeatist response to the Goodman paradox will be that none of the Carnap-Hintikka style systems of inductive logic allows for time-ordering (Putnam 1963). Therefore, the Goodman paradox, does not even arise in the systems, in which it is not possible to express any grue-like predicates. This is not, however, the kind of response I will advance here, for it clearly was Carnap's initial intention to develop the systems to be able to represent order, and time-ordering in particular.[42] The important point is the following: suppose we have observed n individuals. We want then to make a prediction concerning the next individual $n+1$ on the grounds of the evidence we have after having observed n individuals. Goodman's point is that any "projection" onto the future on the basis of the evidence of the past is as good as any other, on syntactical and semantical grounds alone. What, then, will the prediction look like in order to acknowledge the point Goodman advances? In the first instance, the way the upcoming evidence is accommodated within the system of inductive logic will not be favorable to any possible projection. That the acquisition of the upcoming evidence treats, indeed, all possible projections on a par is warranted by what Carnap calls "symmetry conditions".

In other words, what we acquire from evidence is the frequency vector, which is indifferent with respect to individuals observed and the predicates applied to state the results of the observation. However, Goodman's grue-like predicates do not satisfy the symmetry conditions. In effect, the grue-like predicates that we read off from the evidence force us to impose asymmetries between different projections. Goodman seems to force a bias into the mechanism of evidence acquisition—a bias towards some kinds of the possible projections. It is not, however, the acquisition of new evidence that will be biased. Carnap, instead, proposes that what is to be biased is the way that grue-like and non-grue-like predicates affect the degree of confirmation. If we take, for instance, the λ-system, its characteristic function may be presented as a weighted mean of the "empirical" and "linguistic" factors:

$$\frac{s_i}{s}\cdot\left(\frac{s}{s+\lambda}\right)+\frac{1}{\kappa}\cdot\left(\frac{\lambda}{s+\lambda}\right)$$

Suppose now that we have two language systems, which are totally alike except that one of them, call it L_S, has the "Grue" predicate in place of "Green"; call the latter language system L_R. It holds then that $\kappa_R = \kappa_S$, $s_R = s_S$, and s is the same for L_R and L_S. What is still to be determined in a particular context of application, however, is the value of λ_R and λ_S. The context of application, as is persistently emphasized by Carnap, is not a matter of the inductive *logic* alone, but it is a matter of *methodology* of inductive logic instead. A similar point is advocated by Jeffrey, who proposes to redefine projectibility in semantical terms, and then points out that "the impact of Goodman's *Query* on Carnapian confirmation theory is that questions of projectibility can be dealt with quite handily in that theory, once it is placed squarely on the semantical footing it requires" (Jeffrey 1966/1997, 86). In other words, it is the context of application, in which the pragmatic concepts—that Goodman worries so much about—find their suitable place. In fact, it is—for one—the pragmatic concept of the total evidence that will prove essential in assigning a numerical value to λ_R and λ_S. And if we are to make a prediction in L_R and L_S, then total evidence will definitely be different—insofar as it contains the same background knowledge that tells us that grue-like predictions and grue-like hypotheses are counterintuitive—and appropriately will differ the values of λ_R and λ_S. Hence, it is possible that $c(R(a_{n+1}) \mid e_R)$ will be close to 1, while $c(S(a_{n+1}) \mid e_S)$ will approach 0. A similar point trivially holds with respect to general hypotheses in the α-λ continuum.[43]

The point I propounded here can also be extended to cover the modified versions of the systems, which are capable of representing time-ordering. The general strategy outlined by Skyrms in (Skyrms 1994, 247-62) turns on the explanation of projectibility in terms of symmetry of structures, especially in terms of partial exchangeability[44] applied to Markov chains and ergodic theories.

4. The Structural Reliabilist Redefinition of the Bayesian Objective-Subjective Dichotomy

If we agree that Carnap advances a standard empiricist version of foundationalism—like the one in (Ayer 1955) or (Lewis 1946)—then, I admit, there are severe criticisms to be leveled against this position. Apparently though, it would be more appropriate to label Carnap "probabilistic rationalist" rather than empiricist (Jeffrey 1989, 234; 1991, 259; Sarkar 1992b, 8-11).[45] This is because various assumptions which Carnap used in the construction simply do not fit the criteria of objectivity preferred by empiricists. The empiricist expectation concerning inductive logic is that by means of mathematics we extract "probabilities" from innocent data sets. The task of inductive logic would then be to cast different realizations of the procedure of extracting probabilities from data as a homogeneous pattern. Thus the epistemic role of "logic" is nullified, except for a slight epistemic advantage of precision, systematization and apparent simplicity. For logic has no role in objectifying what is taken from the data sets.[46]

Thus among the criteria proposed by Salmon (Salmon 1990b, 184-186) to objectively approach the assessment of priors assigned to hypotheses logic has virtually no epistemic role to play. The pragmatic criteria, e.g. the fact that a non-scientist raised a hypothesis, leave no room for logic at all. Subsequently the formal criteria of consistency of a hypothesis with respect to its components and "accepted" laws and theories are set forth. On the grounds of Salmon's dismissal of partial implication (Salmon 1967b), the role for the formal criteria in setting up priors is minimal, to wit, they only constrain the applicability of a prior ascription to a given hypothesis. Finally, there are material criteria: simplicity, symmetry and analogy. These, although Salmon is silent on them, might be expected to be specified by logic. In either case, however, there is no epistemic role for logic, which only attributes the epistemological significance to the already recognized precise criteria. In fact, empiricists, like Salmon, are forced to advance the claim that inference merely transmits, but by no means accomplishes justification (Audi 1998, 159).

The empiricist worry with respect to logic's epistemic role is worded by Howson and Urbach thus: "we take logic to be essentially noncommittal on substantive matters" (Howson and Urbach 1993, 71-72).[47] However, they mistakenly identify the target, i.e. the analytical nature of Carnap's inductive logic. In these terms the latter is analytical, while also "committal" on epistemic matters.[48] For according to Carnap's structural reliabilist position it only makes sense to pose epistemic questions when a language embedded structure is already determined, which provides the whole semantical set-up necessary and sufficient to determine the epistemic status of a given hypothesis within the language embedded structure (Carnap 1950/1956). In setting forth the role of logic as epistemological means Carnap drew upon Kant's idea that it is primarily the task of logic to provide "the forms of concepts, statements, and inferences, which are then applicable everywhere" (Carnap 1963a, 12-13). In particular the basic idea behind this *structural* approach to objectivity is

> that we can define each scientific concept uniquely by looking at its purely structural relations to all other scientific concepts—only by embedding the concepts into this holistic system can we hope that logical structure will single out relations uniquely. If successful this program would solve the problem of the possibility of objective knowledge (Richardson 1992, 83).[49]

Bayesian-cum-extra view offers an alternative, subjective, interpretation of inductive logic. We do not know whether de Finetti's accepted Carnap's declaration that his inductive logic is, as the Bayesian-cum-extra view would have it, based on the standard axioms of probability theory, on the one hand, and on the extra axioms, on the other. Nevertheless, the followers of de Finetti seem to be rather skeptical concerning the viability of Carnap's Bayesianism. Although Jeffrey apparently adopts the general Bayesian-cum-extra view on Carnap, he does not hesitate to conclude that the extra axioms are "hopeless" on the subjective Bayesian grounds (Jeffrey 1986b, 54), for it narrows the applicability of Carnap's system by, for example, Johnson's sufficientness postulate or exchangeability. Jeffrey seems to be the only one to try to come to grips with this apparent incoherence in Carnap's views. He identifies two senses of "logical" in Carnap's inductive logic. First, it is equivalent to "nonfactual"; second, it means "in agreement with our inductive intuitions as they will exist at the time when the program has been carried out" (Jeffrey 1973, 301).

According to Jeffrey, the process of developing inductive logic from *Logical Foundations of Probability* on induces the development of inductive intuitions, and vice versa. From the subjectivist's perspective, Jeffrey admits, the differences between the confirmation functions of the λ-system, which seem to reflect differences in the individual psychology of epistemic agents, do not undermine the general unity, which seems to reflect general psychology, imposed by the characteristic function G (see definition 1.17 on page 24) on the confirmation functions. The individual differences are, however, swamped by experience, and thus they are only reflected as the argument e of $c(.\,|\,.)$ specifying experience of a given agent, but they do not affect the function itself—"inductive temperament" as Jeffrey calls it.

The important conclusion for the present argument is that Carnap at times "sounds Kantian: one's inductive temperament is *evoked* but not *determined* by one's experience" (Jeffrey 1973, 303), i.e. one's experience "evokes", but does not "determine" the *c*-function. Thus we approach the rationale for a subjectivist not to accept Carnap's extra axioms. For the *c*-function imposes "rationality" upon the empirical agent's beliefs.[50] According to the structural reliabilist theory of justification it is only logic in the form of language embedded structures that imposes epistemic canons on empirical agents. What is unacceptable to Jeffrey is that this way of establishing epistemic canons is insufficient to determine the epistemic status of knowledge states of a given agent. It should be noted, however, that the further condition Jeffrey proposes, i.e. the epistemic responsibility of an agent for collecting evidence, is epistemically irrelevant from Carnap's minimalist standpoint on justification.

Jeffrey's epistemological analysis of Carnap's position, as much as it goes along the lines of structural reliabilism, answers two worries of Salmon's: that it is arbitrary to justify the axioms of inductive logic by mere "inductive intuition" as well as the problem "How can statements that say nothing about any matters of fact serve as a 'Guide to life'?".[51] The answer is in much the same vein as before: logic is committal on epistemic matters, and the only access we have, fallible and vague as it may appear after all, is via inductive intuition (Carnap 1947; 141).

If we suspend the epistemological perspective for a while, the Bayesian-cum-extra reading of Carnap's systems—as shown in detail by Skyrms—seems to promise most in terms of eliminating the limitations concerning the application of the systems of inductive logic. Indeed, it offers a comprehensive accommodation of Carnap's inductive logic, the intended developments included, within the subjective Bayesian statistics.[52] In the classical Bayesian view, Carnap's methods correspond to multinomial sampling with Dirichlet priors, from symmetrical to uniform.[53] Skyrms via Johnson's sufficientness postulate obtains a purely subjective Bayesian reading of Carnap's methods as "strong symmetries in a predictor's degrees of belief" (Skyrms 1996a, 324).

In similar vein Lehrer admits that he regards

> the logical theory of probability as providing some routine ways of formulating subjective probabilities. The logical theorist, like Carnap and Hintikka, I view as making the very important contribution of offering us a variety of different ways of assigning probabilities subjectively. This is a notable contribution. A subjectivist must find some method for expressing a probability assignment that is *coherent* (Lehrer 1981a, 44-45).

The Bayesian view is strongly compelling for not only are Carnap's continua of *The Continuum of Inductive Methods* and their later developments embedded within subjective Bayesian statistics, but Skyrms also extends them within this context to meet the most challenging problems for Carnap: the value continuum, analogy by similarity and proximity, as well as the problem of capturing order relations within inductive logic (Skyrms 1991; 1993; 1996).

Skyrms is, however, quite explicit on the epistemological drawback of this solution:

> from a subjective point of view none of Johnson's postulates has any *logical* status: they merely serve to characterize cases in which degrees of belief have certain interesting and computationally tractable symmetries (Skyrms 1991, 456).

This conclusion would undermine, however, the whole epistemological program of providing the structural reliabilist theory of justification in the form of the systems of inductive logic. For Carnap not only intended to establish the foundations of statistics similar to those of mathematics provided by deductive logic (Salmon 1967a, 726); he also aimed at a coherent *epistemological* foundation of inductive inferences. Thus, in the structural reliabilist view, inductive and deductive logic within a specified language embedded structure are to constitute the exhaustive set of epistemic canons against which to measure the justificatory merits of any scientific sentence. Skyrms is quite aware that once we agree to adopt Carnap's inductive logic within Bayesian statistics the epistemological foundational plans of *Logical Foundations of Probability* are lost. From this perspective Carnap seems to be cast in a trade-off: on the one hand epistemological assumptions lead to computational methods of a very limited area of application (see page 77), whereas on the other, more relaxed computational methods are devoid of epistemological foundation.

This epistemological inadequacy of the standard Bayesian reading of Carnap's and Hintikka's systems of inductive logic is important for Bayesianism itself, as I will argue. A fundamental epistemological dispute within Bayesianism concerns the grounds on which a prior distribution is to be based.[54] Extreme *subjectivist* Bayesians claim that the only basis to ground a prior distribution are personal probabilities, i.e. someone's degrees of belief which satisfy the standard axioms of probability (see page 7). Extreme *objective* Bayesians, on the other hand, claim that a prior distribution must be grounded independently of anyone's degrees of belief, e.g. based on the structure of the available data or the physical chance set-up.[55] Suppose, for instance, that there is a wheel of fortune partitioned into n differently colored fields, one of which is labeled "bankrupt". What is the prior probability, i.e. the probability prior to any spinning of this wheel, that after the first rotation of the wheel it will not stop at the field bankrupt? The extreme subjectivist Bayesian will claim that one is allowed to assign any numerical value from the closed interval [0, 1] provided that the overall distribution on the outcomes of the circulation of the wheel is coherent. Any other constraint beyond one's subjective degrees of belief will be arbitrary and unjustified.[56] The extreme objective Bayesian, on the other hand, will claim that the probability is an output of a function of provided information, and at no point is reference made to subjective degrees of belief.[57]

According to the Bayesian reading of the systems of inductive logic, Carnap and Hintikka endorse the extreme objective Bayesianism because they make a prior distribution depend on the logical structure of the available information. Festa, however, adjusted this wide-spread view to include elements of both extremes of the objective-subjective dichotomy.[58] On this account a prior distribution expresses a subjective uncertainty, which is then confronted with the outcomes of successive spins of the wheel. And these outcomes are determined by the physical chance mechanism that the wheel embeds. Hence, some of the systems of inductive logic of Carnap and Hintikka correspond to multinomial sampling with Dirichlet priors. Skyrms—via Zabell's theorem on equivalence between sufficientness and exchangeability, on the one hand, and Carnap's principles underlying his inductive logic, on the other—takes the Carnap-Hintikka style systems of inductive logic, in particular the γ-λ-system, to the subjectivist extreme. Sufficientness and exchangeability would thus represent strong symmetries among the predictor's beliefs without any recourse to objective chance set-up.

All these attempts to locate Carnap's and Hintikka's systems of inductive logic within the objective-subjective Bayesian dichotomy fail, however, insofar as the standard Bayesian reading of Carnap's and Hintikka's systems fails, as I argued at the beginning of this section. An explanation for this failure might be that Carnap and Hintikka[59] project a hitherto unknown position reconciling the two extremes. Contrary to appearances, Lewis's *Principal Principle* emerges as a good approximation of how to understand Carnap's own approach to the objective-subjective dichotomy[60] (cf. Howson 1995, 26). At the outset Lewis recognizes the dichotomy and acknowledges that "Carnap did well to distinguish two concepts of probability, insisting that both were legitimate and useful and that neither was at fault because it was not the other" (Lewis 1980, 263). Though, Carnap's concepts of "degree of confirmation" and "relative frequency in the long run" are subsumed by, accordingly, "credence or degree of belief", and "chance or propensity". This discrepancy, however, could safely be ignored here (cf. Lewis 1994/1999, 236). What is pertinent to the task at hand is that Lewis seeks to contrive a straightforward and intuitively plausible rule which will comprise elements of both extremes of the dichotomy. The rule in question is called by Lewis, appropriately enough, the Principal Principle:

(PP) Let C be any reasonable initial credence function. Let t be any time. Let x be any real number in the unit interval. Let X be the proposition that the chance, at time t, of A's holding equals x. Let E be any proposition compatible with X that is admissible at time t. Then $C(A \mid XE) = x$ (Lewis 1980, 266).[61]

In his writings on probability Carnap persistently distinguished the logical concept of probability, i.e. "the degree of confirmation", from the empirical concept of probability, i.e. "the relative frequency". Lewis in the Principal Principle tries to overcome this dichotomy, replacing, though, Carnap's initial "degree of confirmation" with "credence", and "the relative frequency" with "chance" (Lewis 1980, 263-64).[62]

Such a rule, or so it would seem, is at variance with Carnap's discrimination between the two extremes of the Bayesian objective-subjective distinction. A minute contemplation of Carnap's characteristic function G (definition 1.17), to take just one example, would suffice to conclude that under appropriate substitution for parameters this formula will resemble Lewis's Principal Principle:

$$c(h \mid e \wedge j) = \mathrm{G}(s, s_i, \kappa) = (1-\zeta)\cdot(s_i / s);$$

where ζ equals to, or approaches, 0.

The total evidence $e \wedge j$ is the best indication of the objective probability of obtaining s_i successes in s experiments, and—on the assumption that the language of science approximates its mature status—this sort of evidence would suffice to determine the degree of confirmation of the hypothesis h under investigation.

Many criticisms are leveled against the Principal Principle in the above cited formulation. Howson, for instance, is ready to acknowledge the principle itself, yet remains skeptical as regards chance. For the Principal Principle fails, or so he claims, to refer to a purportedly mind-independent chance; it rather appeals to a subjective counterpart of chance, i.e. to a certain invariance of subjective beliefs, hence: "the chance of A is that quantity which makes our warranted degree of belief in A constant in just the factual circumstances cited" (Howson 1995, 26). Indeed, it might seem that Lewis is motivated by Hume's skepticism: "Though there be no such thing as *Chance* in the world; our ignorance of the real causes of any event has the same influence on the understanding, and begets a like species of belief or opinion" (Hume 1739/1967, 60). A similar view is advocated by Jeffrey, who follows de Finetti in claiming that chances are like secondary qualities for J. Locke (Jeffrey 1997).[63]

A different concern is raised by Salmon. He recognizes not only the principle, but – in contrast to Howson – also the twofold nature of probability. Still, the problem remains "who is in the driver's seat, subjective credence or objective chance?" (Salmon 1988, 19). Let me explain Salmon's point in terms of "inputs" and "outputs". The point Salmon makes seems to be that the subjective and objective interpretation of the Principal Principle hangs on what the input and output arguments of the principle are. If X and E are inputs, and x is an output, then the Principal Principle expresses, according to Salmon, one of the most elementary requirements of dynamic rationality. The requirement is that one's degrees of belief, or one's "credence function", has to be updated conditional on the upcoming evidence, which reports the new empirical data. However, if E and x are inputs, and X is an output of the Principal Principle, then, Salmon claims, we face a purely subjectivist interpretation of the principle, according to which objective chances are determined by one's credence function. In the wheel of fortune example this is tantamount to acknowledging that the objective chance of the wheel stopping at the field "bankrupt" is determined by one's credence function, provided no defeating evidence is available.

From the perspective of the argument I develop it is, at the moment, irrelevant whether Salmon's criticism of the subjective interpretation of the Principal Principle is correct or not. For Carnap intended the logical concept of probability to be an estimate of the relative frequency, and similarly Lewis intends to determine objective probabilities, i.e. chances, within the subjective framework. Moreover, both Carnap and Lewis intend to define the concept of probability, which will serve as a "guide to life".

Consider first the problem raised by Howson. If he is right that "chance" in the Principal Principle is in fact its subjective counterpart in disguise, then, of course, by no means can the principle be an approximation of the solution to the Bayesian objective-subjective dichotomy propounded in the form of inductive logic. And the reason why can only be acknowledged if we take into account the structural reliabilist position embraced in the systems of inductive logic. In the structural reliabilist view the epistemic status of a given sentence is never a matter of contingency; for it is conferred upon the sentence in question by the logical structure in which the sentence resides. Therefore, in the structural reliabilist view, it is impossible to determine objective chance merely on the grounds of an empirical agent's credence function; such a function is susceptible to empirical investigations, and does not influence the epistemic status of the sentences upon which it operates. Moreover, the credence function of an individual is not determined *a priori*, but it is developed across the individual's lifespan. By no means, then, can a credence function serve to determine or estimate objective chances. It is the other way round, a credence function in order to be rational must itself acknowledge the relevant objective chances. If not, then such a credence function will lead to irrational behavior on the part of the individual who embraces it.

In Lewis's own formulation, however, chance is not – as Howson would have it - credence in disguise:

> chance is neither anyone's actual credence nor the credence warranted by our total available evidence. If there were no believers, or if our total evidence came from misleadingly unrepresentative samples, that wouldn't affect chance in any way (Lewis 1994/1999, 227).[64]

Even if we contend that there are in fact two aspects of probability, then we still need to face the worry raised by Salmon, namely how to make sense of the idea that rational credence determines objective chance. Lewis vehemently protest against this idea:

> Despite appearances and the odd metaphor, this is not epistemology! You're welcome to spot an analogy, but I insist that I am not talking about how evidence determines what's reasonable to believe about laws and chances. Rather, I'm talking about how nature – the Humean arrangement of qualities – determines what's true about the laws and chances. Whether there are any believers living in the lawful and chancy world has nothing to do with it (Lewis 1994/1999, 236).

What still remains unanswered though is Howson's objection that Lewis's Principal Principle and the definition of inductive methods within the Carnap-Hintikka style systems of inductive logic are plagued by the same sort of arbitrariness. Therefore, rational credence, as referred to in the Principal Principle, may only allow for a casual identification with someone's actual degrees of belief. Salmon's worry, it would seem then, is based on a misunderstanding. Still, it would make no difference – as Howson argues – whether we appeal to rational or actual credence, unless it is apparent that the former is not plagued by the same arbitrariness as the latter. Lewis acknowledged the point raised by Howson and modified the Principal Principle. In the new version the principle refers to chance conditional on the total evidence up to time t rather than to unconditional chance (Lewis 1994/1999).

Similarly to the new version of the Principal Principle, the parameters in the systems of inductive logic, such as α, λ, or γ, acquire numerical values only in the context of application, and there are no definite rules on how to accomplish this, in the same way as there are no definite rules on how to provide a credence function in Lewis's account. Actual credence function is a coherent measure of the degrees of beliefs of an empirical cognitive agent. So the credence functions may vary in as much as individuals differ in their degrees of beliefs, provided these are coherent. Neither for Carnap nor for

Hintikka is the choice of the value of a parameter occurring in a system of inductive logic a contingent matter. Roughly, they assume that for a given domain of discourse there is an *a priori* categorization, which is more and more closely approximated by the present development of science. Even if science never reaches the final stage when the basic scientific concepts match the *a priori* categorization, it is still possible that the values of the parameters are set to adequately measure the present state of scientific development. Thus, the parameters are indeterminate and arbitrary insofar as the present state of science is distant from the *a priori* categorization. And the arbitrariness in question diminishes as science progresses. R. Black claims that it is possible to provide a similar interpretation of the Principal Principle as soon as we agree that "the world has got populated with entities which look like objectivised projections of our inductive practices" (Black 1998, 384-85). But it is not at all clear why Lewis might want to embrace this existential claim.

What kind of solution to the problem of the objective-subjective Bayesian dichotomy, Carnap and Hintikka project then? The solution has, I hope, already been made apparent in the preceding discussion. In the first instance, the structural reliabilist view constructed in the systems of inductive logic redefines the concept of objectivity. Primarily, it is the *a priori* conceptual structure that is objective. We already discovered an important part of this conceptual structure in the form of deductive logic, which is based on the concept of logical consequence. Furthermore, Carnap and Hintikka claim to extend the discovery to include also inductive logic as an explication of the concept of the degree of confirmation. The remaining ingredients of the *a priori* conceptual structure have to be unfolded by science, and this, as Carnap admits, is still to be accomplished. Scientific research will be more and more successful in its unfolding the *a priori* conceptual structure as it employs the already discovered *a priori* conceptualization conveyed by deductive and inductive logic. The best illustration is the history of geometry, which succeeded only after embracing logical structure. While the final *a priori* conceptual categorization is far from being known and the "current degree of explicitness of the rules of the language... is still very small" (Carnap 1990, 432), it leaves a rightful place for subjectivity in developing science.

The systems of inductive logic determine the upper and lower boundaries of prior distribution, but the determination of the exact shape of mathematical functions also requires, at *the present state* of scientific development, a subjective assessment of which values of parameters in the systems will be appropriate for a given context of application.[65] Let me return to the wheel of fortune example. What is the prior probability that the wheel will not stop at the field bankrupt? The present development of science does not assure us of a unique objectively determined answer to the question. In the case of the γ-λ-system, for instance, inductive logic determines the two boundaries: $1/\gamma_i \cdot(\alpha)$ and $s_i/s \cdot (1-\alpha)$. The exact values of parameters have still to be assessed. And this assessment cannot be accomplished, except on subjective grounds, i.e. on the grounds of one's acquaintance with the type of the context of application in question as well as one's total evidence. We may only hope that science will at some time develop sufficiently to yield a unique answer, which will rest on the relevant *a priori* categorization.

Inductive logic, however, has already sufficiently developed to cast the relations between factual sentences in *quantitative* form. For any two given factual sentences, i.e. sentences taken from empirical science, we are able to convey the degree of confirmation in the objective form of a numerical measure. And it is due to the incomplete state of scientific knowledge that we still have to rely on subjective assessment, which finally determines which of the possible numerical values from the range determined by the characteristic function are to be picked out in a particular context of application. At present, however, we have no clue as to there being one or more *a priori* conceptual categorizations. It was assumed by some philosophers, most notably by Kant, that there is only one *a priori* conceptual framework. But the emergence of non-Euclidean geometries seems to have undermined this claim. Carnap allows then (see page 88) that it might well be the case that there are more than one *a priori* conceptual categorizations, in fact, each domain of discourse may have its own *a priori* conceptual categorization. The systems of inductive logic are suitably constructed to fit both possible cases.

Kant was quite confident that science of his day reached the ultimate stage; Carnap – being aware of the subsequent accomplishments – is much less so. Therefore, as we have observed, he leaves room for subjectivity in science at the present stage of its progress. Accordingly, as long as there are scientific theories cast in languages allowing for different (number of) primitive predicates (properties), the inductive methods defined in the systems of inductive logic will have to take into account those various possibilities and include the parameters, like κ. Thus, it leads to the conclusion that the ultimate stage of scientific progress will become manifest only when there only be a unified language of science with uniquely determined number of primitive predicates, because at present: "our empirical *language* can only be understood as an incoherent and fragmentary schema of an ideally coherent *language*" (Sellars 1948/1959, 456). It is at this final stage that Lewis's Principal Principle, in the initial form, could be safely derived from the systems of inductive logic. The degree of confirmation of a given hypothesis conditional on total evidence will supervene on nothing but relative frequency. At this advanced stage, however, (the limit of) relative frequency (or chance) will be a consequence of the ultimate theory (or "the best system" to use Lewis's favorite term) rather than an empirical generalization. The supervenience of the degree of confirmation on the total evidence – summing up relative frequency – will only be possible, as Lewis's Principal Principle demonstrates, in the absence inadmissible evidence. There seem to be two sorts of candidates for inadmissible evidence-sentences in the systems of inductive logic (cf. Carnap 1950/1956). First, any evidence-sentence formulated in a language which allows for a different number of primitive predicates than the one in which a hypothesis is cast. But this option is automatically ruled out, especially at the stage when the unified language of science would determine a unique number of primitive predicates. Second, given the unified language, there is still an infinite number of evidence-sentences. And, as Lewis concedes, there is no conclusive argument to the effect that there will only be one evidence-sentence to satisfy plausible—and perhaps structurally supervenient—criteria, such as simplicity, strength and fitness. Unless such a stage of scientific progress is reached and the unique total evidence is picked out, the systems of inductive logic have to accommodate some indeterminacy as they could not accomplish more than science itself.

Notes

1 See (Pollock 1986, 121-22); (Plantinga 1993, 184-208) for a critical discussion.
2 A similar point is advanced in (Hintikka 1971, 335-36); (Hilpinen 1973, 310).
3 It might be argued, however, that any theory of probability, which embraces normalization can only be compatible with a non-subjectivist interpretation, in particular with non-subjectivist Bayesianism; cf. (Cooke 1986, 26).
4 See (Skyrms 1992) for a discussion of how to proceed from finite to countable additivity on B. de Finetti's grounds.
5 For a general discussion see of the concepts of "analyticity" see (Przełęcki and Wójcicki 1969/1977).
6 The empiricist grounds for W. C. Salmon's criticism to follow are challenged by L. BonJour; cf. (BonJour 1998a, 32-58).
7 A still stronger claim is that the required justification needs to be grounded in lawlike dependencies among events reported and predicted; cf. (Goldman 1988, 26-27). This view was incorporated into the systems of inductive logic by J. Hintikka (Hintikka 1966), J. Pietarinen (Pietarinen 1972), I. Niiniluoto (Niiniluoto 1981b) and R. Festa (1993).
8 Cf. (Salmon 1967a, 736); (Hacking 1968, 53); (Howson 1995, 4). For a critical discussion of this and alternative ways to justify the standard axioms of probability theory see (Shimony 1988); (Maher 1997).

9 Cf. (Carnap 1963, 978); (Salmon 1967a, 735-38; 1967b, 79); (Kuipers 1983); (Hintikka 1991b, 130-31).

10 (Kuipers 1976a, 88-89); (Kuipers 1978b, 38-70); (Jeffrey 1981). For a discussion of the difficulties to overcome see (Łoś 1960, 110-13); (Humphreys 1995/1996, 119).

11 I shall not study here the question of whether the epistemic principles in the other systems are homogeneous with those underlying Carnap's system.

12 A. C. Grayling perceives the transition from minimalist theory to mentalistic theory of justification as one of the essential epistemological problems, which has its source in the cognitive limitations of empirical cognitive agents—"the finitary predicament"—cf. (Grayling 1995).

13 For an alternative semantical approach to confirmation as a "cognate of the notion of truth" see (Wójcicki 1976, 14-15).

14 R. Chisholm points out that in a context of application Carnapian concept of total evidence will inevitably be relativized to total evidence of a given epistemic agent. Therefore, the initial definition of *confirmation* provided by Carnap has to be suitably refined in order to be useful for the definition of knowledge; cf. (Chisholm 1966/1977, 72; 82-83). Such a modification, which turns on *S*'s computational capacities, see (Swinburne 1997, 158).

15 T. Williamson defends the view that knowledge and evidence are equivalent; cf. (Williamson 1997; 1998, 100-102). For a criticism see (Maher 1996).

16 A similar argument is advocated by (Holland et al. 1986, 2-5).

17 R. Hilpinen in (Hilpinen 1971) gives good grounds for skepticism regarding these plans.

18 In the first instance, I will omit those paradoxes, which are trivially resolved within the Carnap-Hintikka style systems of inductive logic.

19 I. Hacking, however, claims that this kind of paradox is historically and conceptually prior to Hume's problem of induction; cf. (Hacking 1993/1997); see also (Kaplan 1998, 745-51).

20 A similar point is advanced by E. Sosa; cf. (Sosa 1991, 127).

21 For a discussion see (Toulmin 1966, 129-41).

22 In representing the argument I follow (Suppes 1966a, 198-200); (Mortimer 1982, 11.4) rather than (Hempel 1962, 30-36).

23 Achinstein in (Achinstein 1978) set forth several difficulties against Carnap's concept of evidence; for a critical assessment of Achinstein's arguments see (Kronz 1992) and for Achinstein's response to them see (Achinstein 1992). See also (Suppes 1966b, 50); (Salmon 1981, 272).

24 See also (Michalos 1971b, 55-59); (Pollock 1973). For a survey of recent discussions see (Maher 1999).

25 For a criticism of this solution see (Black 1966, 195-97).

26 J. Agassi raised the criticism that this is merely a linguistic trick, not a solution; cf. (Agassi 1966, 400).

27 See (Achinstein 1992, 173) for a list of proposals of similar solutions of the paradox. For Bayesian solutions of the paradox see (Howson and Urbach 1993, 126-30); (Achinstein 1993, 173-74).

28 The flaws were pointed out by Black (Black 1966, 195-97), and analyzed in detail by Mortimer (Mortimer 1982, 187-89).

29 This limitation can easily be removed if we define $P(p) = c(p \mid t)$.

30 This was the point of the counter-argument leveled by Black; cf. (Black 1966, 195-97).

31 An approach to resolve the ravens paradox on the grounds of the Carnap-Hintikka style systems of inductive logic which is allegedly adjusted to take into account the zero-confirmation problem see (Humburg 1986).

32 Thirty years later or so, J. Earman questioned the grounds for Suppes's optimism (Earman 1992, Ch. 6).

33 This is precisely for that reason that Hintikka's solution to the ravens paradox based on the presupposition view of induction surpasses the structural perspective adopted here; cf. (Hintikka 1969, 28-34).

34 For a discussion see (Skyrms 1966a, 57-69; 1966b; 1972). For an extensive annotated bibliography on the Goodman paradox see (Stalker 1994, 281-457).

35 A similar point is advocated by S. S. Shoemaker, but his solution of the paradox involves the presupposition view on induction; cf. (Shoemaker 1980); and (Swinburne 1980) for a criticism. For a defense of Carnap's claim see (Salmon 1963b, 257-61; 1969; 1974).

36 The same point, *mutatis mutandis*, can be easily shown to hold for Hintikka's systems. I limit my discussion basically to Carnap's systems at the moment, for it was Goodman's initial target.

37 It follows from Carnap's symmetry condition on predicates as well; (see page 20).

38 In addition, Carnap and Hempel objected that this example violates the principle of total evidence; cf. (Carnap 1947/1997a, 8-9); (Hempel 1962, 150).

39 Cf. (Goodman 1947/1997a). For a response see (Carnap 1947/1997b). For a defense of Carnap's original objection see also (Kahane 1965); (Goldstick 1989); and (Weir 1995) for criticism.

40 For a solution alternative to the one I develop here see (Dubucs 1993a, 93-96).

41 Although H. Kahane attempts to support the view that even on the grounds of Carnap's inductive logic alone it is possible to formulate hypotheses, which are not projectable; cf. (Kahane 1965). This attempt has been dismissed by R. J. Ackerman, who showed how to make Kahane's hypotheses projectable; see (Ackerman 1966, 71-72).

42 Such a development was already accomplished on Bayesian grounds, however, by Skyrms in (Skyrms 1991); for the details see the fourth section of the first chapter of the present book.

43 The alternative solution Hintikka and his co-workers propound (Hintikka 1969); (Pietarinen 1972) relies on thoroughly different sort of evidence which surpasses the structural perspective adopted here. Therefore I relegate the study of this solution to a separate study.

44 For the concept of *partial exchangeability* see (De Finetti 1980, 194); (von Plato 1981a; 1981b).

45 For a defense of a similar view concerning J. Keynes see (O'Donnell 1990).

46 For a criticism of the objective interpretation of the concept of probability see e.g. (De Finetti 1982, 392-94).

47 A similar objection is raised in (Fine 1973, 203).

48 Carnap's conception of *inductive* logic is particularly difficult to swallow for an empiricist, because it conflicts with the empiricist conception of analyticity; cf. (BonJour 1993b, 394); (BonJour 1998a, 30-32).

49 Nevertheless, since (Carnap 1962) he seems to have resigned from this epistemological role of logic, while assuming also that the essential concept of degree of confirmation is no longer a purely logical concept, but is primarily a decision-theoretic one; cf. (Carnap 1963b, 67).

50 The priority in attributing the epistemic merit to a belief-generating or justification-conferring mechanism seems to be a generic feature of reliabilist epistemologies; cf. (Sosa 1991, 210-11).
51 (Salmon 1967b, 76-78). This claim R. Carnap advances first in (Carnap 1947, 141).
52 The subjective reading on the Bayesian view is not the only one; an objective alternative is presented by Festa, who appends pragmatic assumptions to the initial version of Carnap's theory (Festa 1993).
53 Cf. (Kuipers 1978b; Festa 1993; Skyrms 1996; Festa 1999b).
54 For a comprehensive overview of the debate see (Festa 1993, 75-101); (Maher 1996); (Festa 1999a, 76-77).
55 See e.g. (Rosenkrantz 1981, IV and IV*); (Skyrms 1993e, 374-76); (Howson 1995). For a criticism of this view see (Galavotti 1989, 244-55); (Howson 1995, 12-13).
56 For a characterization of the subjective interpretation of probability see e.g. (Lehrer 1981a, 46-47); (Humphreys 1998, 703-704).
57 Cf. (Humphreys 1998, 702-704). Fine, however, claims that the two ways of determining priors might be complementary rather than mutually exclusive; cf. (Fine 1973, 109). Kyburg claims to the contrary that exclusively one interpretation is correct, because collectively they have inconsistent consequences; cf. (Kyburg 1966).
58 There are also alternative attempts to reconcile the two extreme positions. T. Fine, for instance, claims that there are appropriate contexts of application for both (Fine 1973, 109). D. A. Gillies, on the other hand, develops "intersubjective" interpretation of probability; cf. (Gillies 1991).
59 In the remainder of this section I will focus on Carnap's version of inductive logic. For Hintikka and his co-workers propound an alternative standpoint on the objective-subjective dichotomy (Hintikka 1969); (Pietarinen 1972). The latter programmatically draws upon a kind of evidence which from the structural perspective apparently is inadmissible, and therefore I relegate the discussion of this solution to a separate study.
60 For a similar proposal of "objectification" of subjective probability see (Jeffrey 1965, 190-96; 1977, 217-19; 1986a).
61 As Lewis correctly acknowledges this formulation requires further explanations, and these—as well as criticisms in e.g. (Hall 1994)—lead him to provide a more sophisticated version of the Principal Principle; cf. (Lewis 1980, 277); (Lewis 1994/1999, 476). Since my argument does not turn on the simplifications present in the quoted formulation of the principle, I will mainly refer to the less technical formulation of the principle.
62 For a discussion of the concept of *chance* from the perspective of the Carnap-Hintikka style systems of inductive logic see (Swinburne 1971); (Salmon 1979, 198-208); (Niiniluoto 1988b, 98-100).
63 For a criticism of these attempts to embrace "objectification" of probability see (Kyburg 1967); (Salmon 1988).
64 The discussion of Howson's argument to the effect that The Principal Principle is viable only if its according elements are subsumed by their subjective counterparts would surpass the limits of this book.
65 Festa's research on the optimum inductive methods provides a further approximation of the exact shape of the mathematical functions conditional, however, on the available background knowledge concerning the context of inquiry.

Conclusion

The laborious task of fostering a cohesive theory is always a challenge attended to by philosophers. As it appears, Carnap is the author of the two most systematic philosophical constructions propounded in the 20th century. The first one, undertaken in his *Der logische Aufbau der Welt* (Carnap 1928), is regarded as primarily addressing epistemological issues. The other one—*Logical Foundations of Probability*—subjected to scrutiny in this book, appears to share the epistemological concerns of the former. The epistemological approach towards the systems of inductive logic developed by Carnap and the members of the Finnish School, which I adopt throughout this book, resulted in the elaboration of structural reliabilism. I argued at length that formal systems of inductive logic constitute the structural reliabilist theory of justification. I also advanced numerous arguments demonstrating that this theory is an important alternative to the mainstream mentalistic theories of justification as well as arguments discharging the most severe objections leveled against structural reliabilism.

However, it is still an open problem, and indeed a pressing one, whether it is possible to extend the structural reliabilist theory of deductive and non-demonstrative *inferential* of justification to become also a theory of *direct* justification, and accomplish thereby a full-blooded structural reliabilist theory of justification. Alternatively, the version of structural reliabilism developed here could perhaps be elaborated as continuous with the kind of theory of justification which dispenses with direct justification for "nothing can count as a reason for holding a belief except another belief" (Davidson 1983, 426).

A further study aiming at a comprehensive account of structural reliabilism would have to resolve also the issue of how this theory of justification relates to the debate between epistemological externalism and internalism. While analyses of Goldman's process reliabilism would suggest that reliabilism constitutes generically an externalist theory of justification (BonJour 1993a, 133; Alston 1998, 824), there are good reasons to countenance this view with respect to structural reliabilism. Still, to render an adequate account of the relation of structural reliabilism to both externalism and internalism, it would be necessary to account for the arguments which stem from epistemological theories that are opposite to externalism, in general, and which stem from evidentialism, in particular (Feldman and Conee 1985; Cohen 1995, 115; Kawalec 1997c).

The question that a comprehensive appraisal of structural reliabilism would have to face concerns the extent to which this epistemological theory falls within the naturalistic paradigm set forth for epistemology by W. V. Quine:

> Epistemology, or something like it, simply falls into place as a chapter of psychology and hence of natural science. It studies a natural phenomenon, viz., a physical human subject" (Quine 1969, 82).

Of course, this naturalistic paradigm is itself not immune from objections (Siegel 1980; Stroud 1985; Almeder 1990/1993). Again, the analyses of Goldman's reliabilism would suggest that it is a naturalistic theory, indeed, a sophisticated one, which could accommodate normativity. It is claimed, for instance, that Goldman's process reliabilism accords well with evolutionary epistemology (Campbell 1987, 142-43; Campbell and Paller 1989, 247-49). Nevertheless, the naturalistic leanings of process reliabilism, as compellingly demonstrated by Sosa, are themselves debatable (Sosa 1991, 100; Foley 1998, 162; Goldman 1998, 204-205).

Admittedly, the issue of naturalized epistemology is germane to the assessment of structural reliabilism insofar as arguments similar to those put forward against naturalized epistemological theories, could also be advanced against minimalist theories of justification in favor of mentalistic theories of justification (Nagel 1944/1949, 197-203; Stroud 1981; Kim 1988; Rorty 1979; Kitcher 1992; Callebaut 1995b, 35).

The challenge to logical positivism brought up by T. Kuhn seems to have been crucial to the origin and development of a naturalistic philosophy of science and naturalistic epistemology (Lakatos 1971, 108-12; Kuhn 1971, 139; Callebaut 1995a, 3-4; 1995b, 35). Still, it is a pressing question to what extent Kuhn's challenge can be met on Bayesian grounds, and on the grounds of structural reliabilism in particular (Salmon 1970; 1989; 1990b; 1990c; Rosenkrantz 1980, 81; Sintonen 1997a, 1-2). In fact, the very possibility of elaborating the structural reliabilist theory of justification which would be adequate to issues of justification raised in the actual history of science has to be assessed as well (Callebaut 1987, 151; Emmerich and Greeno 1966).

The problem of whether structural reliabilism could also constitute the set of criteria of dynamic rationality, to wit, the criteria of how to change one's beliefs rationally, is a pressing one, if structural reliabilism is to be fully accommodated within Bayesian epistemology. In particular, the problem is pressing for those who claim that Bayesianism in the philosophy of science has replaced Hempel's model of explanation.[1]

The results of the present study are pertinent to the blossoming field of historical studies of the *Aufbau*. Richardson advances the claim that it is precisely the unfolding of Carnap's epistemology in the *Aufbau* that will ground the historically correct overall view of the philosophy of logical empiricists (Richardson 1998, 3). The analogies between the constructions in the *Aufbau* and in *Logical Foundations of Probability* substantiate the conjecture that structural reliabilism might also be illuminating as regards the epistemological position Carnap propounded in the *Aufbau*. From the historical-philosophical perspective it seems that a lot has still to be accomplished in order to appropriately depict how Carnap's philosophy diverges from the traditional logical empiricism advocated by B. Russell. For even Carnap's closest allies, such as H. Feigl and C. G. Hempel, misconceived the historical context of his work.[2]

Perhaps due to this misunderstanding Carnap continues to be influential, but not by the *theses* he propounded, but rather by his *way* of practicing philosophy (Spohn 1991b, 1). Creath claims further that: "prominent contemporary philosophers (Quine and Kuhn) who are thought to have demolished logical empiricism are shown to exhibit central views of the logical empiricists rather than having overthrown them" and it is Carnap who is "undoubtedly a central figure, and very often when people speak of logical positivism they mean Carnap" (Creath 1995/1996, 285, 288). On the other hand, however, M. P. Bradie elucidates an approach to B. Russell's epistemology, which is similar to the structural reliabilist approach advanced here (Bradie 1977).

G. H. von Wright discusses the general cultural context of the philosophical program developed by Carnap throughout his career. Carnap's tendency to render his views in a form of a logical construction is a descendant of the modernist and rationalistic belief in progress. Apart from being social, cultural, and moral, the progress is characteristically scientific. Science develops towards unification. Therefore, as Carnap thought, the task of the philosopher is to construct *the logic of the language* of unified science (Wright 1990, 47). However, the pertinent task of the philosopher set forth by Carnap is primarily, as my study substantiates, to construct *the logic of justification* for unified science: "Logic is no longer merely one philosophical discipline among others ... Logic is the method of philosophizing" (Carnap 1930/1959, 133).

Notes

1 The claim is advocated by W. Salmon, cf. (Salmon 1990a); see also (Skyrms 1987a; 1987b); (Howson 1990a); (Howson 1991a); (Earman and Salmon 1992, 89-95); (Earman 1992); (Howson and Urbach 1993, 115-70); (Skyrms 1993a; 1993d).

2 (Putnam 1963, 761); (Hausman and Wilson 1967); (Achinstein and Barker (eds.) 1969); (Salmon 1974); (Feigl 1975); (Quine 1975, xxiv); (Grünbaum and Salmon (eds.) 1988).

References

Achinstein P., 1978, "Concepts of Evidence", *Mind* 87, 22-45.
Achinstein P., 1992, "The Evidence Against Kronz", *Philosophical Studies* 67, 169-75.
Achinstein P., 1993, "Hempel's Paradox of the Ravens", in Dancy and Sosa (eds.) 1993, 172-74.
Achinstein P., 1997, "On Evidence: A Reply to McGrew", *Analysis* 57, 81-83.
Achinstein P., and Barker S. F. (eds.), 1969, *The Legacy of Logical Positivism*, Baltimore: Hopkins.
Ackerman R. J., 1966, "Discussion: Projecting Unprojectibles", *Philosophy of Science* 33, 70-75.
Adams W. E., 1961, "The Empirical Foundations of Elementary Geometry", in Feigl and Maxwell 1961, 197-212.
Adler J. E., 1980, "Criteria for A Good Inductive Logic", in Cohen and Hesse 1980, 379-405.
Agassi J., 1966, "The Mystery of the Ravens", *Philosophy of Science* 33, 395-402.
Ajdukiewicz K., 1965, "The Problem of Foundation", in K. Ajdukiewicz (ed.) 1965, *The Foundation of Statements and Decisions*, Warszawa: PWN, 1-11.
Almeder R., and Hogg F., 1989, "Reliabilism and Goldman's Theory of Justification", *Philosophia* 19, 165–88.
Almeder R., 1990/1993, "On Naturalizing Epistemology", *American Philosophical Quarterly* 27. Reprinted in Fetzer (ed.) 1993, 451–473.
Alston W., 1989, *Epistemic Justification: Essays in the Theory of Knowledge*, Ithaca: Cornell University Press.
Alston W., 1995a, "Epistemic Warrant as Proper Function", *Philosophy and Phenomenological Research* 55, 397-402.
Alston W., 1995b, "How to Think about Reliability", *Philosophical Topics* 23, 1-29.
Alston W., 1998, "Internalism and Externalism in Epistemology", in Craig (ed.) 1998, Vol. IV, 821-26.
Aristotle, *The Complete Works of Aristotle*. Edited by J. Barnes, Princeton: Princeton University Press 1984, 2 vols.
Armstrong D. M., 1973, *Belief, Truth and Knowledge*, Cambridge: Cambridge University Press.

Audi R., 1998, *Epistemology: A Contemporary Introduction to the Theory of Knowledge*, London: Routledge.
Ayer A. J., 1955, *The Foundations of Empirical Knowledge*, London: Macmillan 1955.
Ayer A. J., 1957, "The Conception of Probability as a Logical Relation", in S. Körner (ed.) 1957, *Observation and Justification*, New York, 12-17.
Ayer A. J. (ed.), 1959, *Logical Positivism*, Glencoe: Free Press.
Barcan Marcus R., 1981, "Modal Logic, Modal Semantics and Their Applications", in Floistad (ed.) 1981, 279-98.
Bar-Hillel Y. (ed.), 1965, *Logic, Methodology, and Philosophy of Science*, Amsterdam: North–Holland.
Bar-Hillel Y., 1968a, "Freudenthal Contra Methodologos", in Lakatos (ed.) 1968b, 17-21.
Bar-Hillel Y., 1968b, "On Alleged Rules of Detachment in Inductive Logic", in Lakatos (ed.) 1968b, 120-28.
Bar-Hillel Y., 1968c, "The Acceptance Syndrome", in Lakatos (ed.) 1968b, 150-161.
Bar-Hillel Y., 1975, "Homage to Carnap", in Hintikka (ed.) 1975b, XXIX-XXX.
Barker S. F., 1957, *Induction and Hypothesis: A Study of the Logic of Confirmation*, Ithaca: Cornell University Press.
Bartha P., Johns R., 2001, "Probability and symmetry", *Philosophy of Science* (*Suppl.*) 68, S109-S122
Batens D., 1975, *Studies in the Logic of Induction and in the Logic of Explanation*, Brugge: De Tempel.
Bayes T., 1763/1958, "An Essay Towards Solving a Problem in the Doctrine of Chances", *Philosophical Transactions of the Royal Society* 53, 370-418; 54, 296-325. Reprinted in *Biometrika* 45, 293-315.
Black M., 1966, "Notes on the "Paradoxes of Confirmation"", in Hintikka and Suppes (eds.) 1966, 175–197.
Black R., 1998, "Chance, Credence, and the Principal Principle", *British Journal for the Philosophy of Science* 49, 371-85.
Bogdan R. J., 1980, "Two Turns in Induction", in Cohen and Hesse (eds.) 1980, 406-415.
Bogdan R. J. (ed.), 1981, *Keith Lehrer*, Dordrecht: Reidel.
Bogdan R. J. (ed.), 1987, *Jaakko Hintikka*, Dordrecht: Reidel.
Bogdan R. J. and Niiniluoto I. (eds.), 1973, *Logic, Language and Probability: A Selection of Papers Contributed to the 4th International Congress for Logic, Methodology, and Philosophy of Science, Bucharest, September 1971*, Dordrecht: Reidel 1973.
Bohnert H. G., 1975a, "Carnap's Logicism", in Hintikka (ed.) 1975b, 183-216.
Bohnert H. G., 1975b, "Homage to Carnap", in Hintikka (ed.) 1975b, XXXI-XLIV.
BonJour L., 1993a, "Externalism/internalism", in Dancy and Sosa (eds.) 1993, 132–136.
BonJour L., 1985, *The Structure of Empirical Knowledge*, Cambridge: Harvard University Press.
BonJour L., 1993b, "Problems of Induction", in Dancy and Sosa (eds.) 1993, 391-395.
BonJour L., 1998a, *Defense of Pure Reason: A Rationalist Account of A Priori Justification*, Cambridge: Cambridge University Press.
BonJour L., 1998b, "Knowledge and Justification, Coherence Theory of", in Craig (ed.) 1998, Vol. V, 253-59.
Bradie M. P., 1977, "The Development of Russell's Structural Postulates", *Philosophy of Science* 44, 441-463.
Braine M. D. S., and O'Brien D. P. (ed.), 1998, *Mental Logic*, Mahwah: LEA.
Brody A. B. (ed.), 1970, *Readings in the Philosophy of Science*, Englewood Cliffs: Prentice-Hall.

Bronk A., 1995, "Filozofia i nauka: problem demarkacji", *Roczniki Filozoficzne* 43, f.1, 181-236.

Bronk A., 1997, "The Axiological and Deontological Dimension of Justification", in Wójcicki (ed.) 1997, 9-30.

Bronk A., 1998a, "Alvina Plantingi koncepcja uzasadniania", in J. Cichoń (ed.), *Ratio et revelatio. Z refleksji filozoficzno-teologicznych*, Opole: WTUO, 71-81.

Bronk A., 1998b, *Zrozumieć świat współczesny*, Lublin: TN KUL.

Buck R. C., and Cohen R. S. (eds.), 1971 , *PSA 1970. In memory of Rudolf Carnap*, Dordrecht: Reidel.

Burks A. W., 1953, "The Presupposition Theory of Induction", *Philosophy of Science* 20, 177-97.

Burks A. W., 1963, "On the Significance of Carnap's System of Inductive Logic", in Schilpp (ed.) 1963, 739-59.

Byrne E. F., 1968, *Probability and Opinion: A Study in the Medieval Presuppositions of Post-medieval Theories of Probability*, The Hague: Nijhoff.

Callebaut W., 1987, "Post–Positivist Views of Scientific Explanation", in J. Duchene, G. Wunsch, and E. Vilquin (eds.) 1987, *Explanation in the Social Sciences: The Search for Causes in Demography*, Bruxelles: Ciaco, 141–196.

Callebaut W., 1995, "Science Dynamics: The Difficult Birth of a Metascience", in H. Stachowiak (ed.) 1995, *Pragmatik: Handbuch Pragmatischen Denkens*, Vol. 5, Hamburg: Felix Meiner, 1–29.

Callebaut W., 1995, "The Future of Naturalistic Philosophy of Science", *Ludus Vitalis* 3, 19–51.

Campbell D. T., 1986, "Science's Social System of Validity–Enhancing Collective Belief Change and the Problems of the Social Sciences", in D. W. Fiske, and R. A. Shweder (eds.) 1986, *Metatheory in Social Science*, Chicago: Chicago University Press, 108–35.

Campbell D. T., 1987, "Selection Theory and the Sociology of Scientific Validity", in W. Callebaut, and R. Pinxten (eds.) 1987, *Evolutionary Epistemology*, Dordrecht: Reidel, 139–58.

Campbell D. T., and Paller B. T., 1989, "Extending Evolutionary Epistemology to "Justifying" Scientific Beliefs", in K. Hahlweg, C. A. Hooker (eds.) 1989, *Issues in Evolutionary Epistemology*, Albany: SUNY University Press, 231–57.

Carnap R., 1928, *Der logische Aufbau der Welt*, Berlin: Weltkreis.

Carnap R., 1930/1959, "The Old and the New Logic", *Erkenntnis* 1. Translated from German and reprinted in Ayer (ed.) 1959, 133-46.

Carnap R., 1936/1949, "Wahrheit und Bewaehrung", in *Actes du Congres International de Philosophie Scientifique* 1936, Paris: Hermann. Translated as "Truth and Confirmation" by H. Feigl, in Feigl and Sellars (eds.) 1949, Readings in Philosophical Analysis, New York: Appleton–Century–Crofts, 119-127.

Carnap R., 1938/1949, "Logical Foundations of the Unity of Science", in O. Neurath et al. (eds.), 1938, International Encyclopedia of Unified Science, Vol. 1. Reprinted in Feigl and Sellars (eds.) 1949, 408-423.

Carnap R., 1945/1970, "On Inductive Logic", *Philosophy of Science* 12, 72-97.

Carnap R., 1945, "The Two Concepts of Probability", *Philosophy and Phenomenological Research* 5, 513-532.

Carnap R., 1946, "Remarks on Induction and Truth", *Philosophy and Phenomenological Research* 6, 590-601.

Carnap R., 1947/1997a, "On the Application of Inductive Logic", *Philosophy and Phenomenological Research* 8, 133-147. Reprinted in Elgin (ed.) 1997, 1-15.

Carnap R., 1947, "Probability As A Guide in Life", *The Journal of Philosophy* 44, 141-148.

Carnap R., 1947/1997b, "Reply to Nelson Goodman's 'On Infirmities of Confirmation-theory'", *Philosophy and Phenomenological Research* 8, 461-462. Reprinted in Elgin (ed.) 1997, 21-22.

Carnap R., 1950/1956, "Empiricism, Semantics, and Ontology", *Revue Internationale de Philosophie* 4, 20-40. Reprinted in Carnap 1956, 205-221.

Carnap R., 1950, *The Logical Foundations of Probability*, Chicago: University of Chicago Press.

Carnap R., 1951a, *The Nature and Application of Inductive Logic, Consisting of Six Sections from Logical Foundations of Probability*, Chicago: University of Chicago Press.

Carnap R., 1951b, "The Problem of Relations in Inductive Logic", *Philosophical Review* 2, 75-80.

Carnap R., 1952a, "Meaning Postulates", in *Philosophical Studies* 3, 65-73.

Carnap R., 1952b, *The Continuum of Inductive Methods*, Chicago: University of Chicago Press.

Carnap R., 1953a, "Inductive Logic and Science", *Proceedings of the American Academy of Arts and Sciences* 80, 189-197.

Carnap R., 1953b, "On the Comparative Concept of Confirmation", *British Journal for the Philosophy of Science* 3, 311-318.

Carnap R., 1953c, "Remarks to Kemeny's Paper, "A Contribution to Inductive Logic"", *Philosophy and Phenomenological Research* 13, 375-376.

Carnap R., 1953d, "What is Probability?", *Scientific American* 189, 128-138.

Carnap R., 1955/1956, "On Some Concepts of Pragmatics", *Philosophical Studies* 6, 89-91. Reprinted in Carnap 1956, 248-50.

Carnap R., 1955, *Statistical and Inductive Probability*, Brooklyn: Galois Institute of Mathematics and Art.

Carnap R., 1956, *Meaning and Necessity: A Study in Semantics and Modal Logic*. Second edition, Chicago: University of Chicago Press.

Carnap R., 1956/1996, "The Methodological Character of Theoretical Concepts", in H. Feigl and M. Scriven (eds.) 1956, *The Foundations of Science and the Concepts of Psychology and Psychoanalysis*, Minneapolis: University of Minnesota Press, 38-76. Reprinted in Sahotra (ed.) 1996, 130-168.

Carnap R., 1962, "The Aim of Inductive Logic", in Nagel, Suppes and Tarski (eds.) 1962, 303–318.

Carnap R., 1963a, "Intellectual Autobiography", in Schilpp (ed.) 1963, 3-84.

Carnap R., 1963b, "Remarks on Probability", *Philosophical Studies* 14, 65-75.

Carnap R., 1963c, "Replies and Systematic Expositions", in Schilpp (ed.) 1963, 859-1013.

Carnap R., 1963d, "Variety, Analogy, and Periodicity in Inductive Logic", *Philosophy of Science* 30, 222-227.

Carnap R., 1968a, "Inductive Logic and Inductive Intuition", in Lakatos (ed.) 1968b, 258-67.

Carnap R., 1968b, "On Rules of Acceptance", in Lakatos (ed.) 1968b, 146-50.

Carnap R., 1968c, "The Concept of Constituent-structure", in Lakatos (ed.) 1968b, 218-20.

Carnap R., 1971a, "A Basic System of Inductive Logic, Part I", in Carnap and Jeffrey (eds.) 1971b, 33–165.

Carnap R., 1971b, "Further Material for the Introduction", in Carnap and Jeffrey (eds.) 1971b, 3-4.

Carnap R., 1971c, "Inductive Logic and Rational Decisions", in Carnap and Jeffrey (eds.) 1971b, 5–31.

Carnap R., 1973, "Notes on Probability and Induction", *Synthese* 25, 269-298.

Carnap R., 1975, "Observation Language and Theoretical Language", in Hintikka (ed.) 1975b, 75-85.

Carnap R., 1977, *Two Essays on Entropy*. Edited by A. Shimony, Berkeley: University of California Press.

Carnap R., 1980, "A Basic System of Inductive Logic, Part 2", in Jeffrey (ed.) 1980b, 7-155.

Carnap R., and Jeffrey R. C., 1971a, "Introduction", in Carnap and Jeffrey (eds.) 1971b, 1–2.

Carnap R., and Jeffrey R. C. (eds.), 1971b, *Studies in Inductive Logic and Probability, Vol. 1*, Berkeley, Los Angeles: University of California Press.

Carnap R., and Stegmüller W., 1959, *Induktive Logik und Wahrscheinlichkeit*, Wien: Springer.

Chatalian G., 1952, "Probability: Inductive versus Deductive", *Philosophical Studies* 3, 49-56.

Chisholm R. M., 1957, *Perceiving: A Philosophical Study*, Ithaca, New York: Cornell University Press.

Chisholm R. M., 1966/1977, *Theory of Knowledge*. Second edition, Englewood Cliffs: Prentice-Hall.

Chisholm R. M., 1982, *The Foundations of Knowing*, Minneapolis: University of Minnesota Press.

Chuaqui R., 1991, *Truth, Possibility, and Probability: New Logical Foundations of Probability and Statistical Inference*, Amsterdam: North-Holland.

Chuaqui R., 1994, "Random Sequences and Hypotheses Tests", in P. Humphreys (ed.), *Patrick Suppes: Scientific Philosopher, Vol. 1*, Dordrecht: Kluwer, 63-85.

Cocchiarella N., 1981, "R. Montague and the Logical Analysis of Language", in Floistad (ed.) 1981, 113-54.

Coffa A., 1974, "Hempel's Ambiguity", *Synthese* 28, 141-63.

Coffa A., 1993, *The Semantic Tradition from Kant to Carnap: to the Vienna Station*, Cambridge: Cambridge University Press.

Cohen L. J., 1973, "A Note on Inductive Logic", *The Journal of Philosophy* 70, 27-40.

Cohen L. J., 1979, "On the Psychology of Prediction: Whose is the Fallacy?", *Cognition* 7, 385-407.

Cohen L. J., 1980, "What Has Inductive Logic to do with Causality", in Cohen and Hesse (eds.) 1980, 156-171.

Cohen L. J., 1993, "What Has Probability to Do with Strength of Belief", in Dubucs (ed.) 1993b, 129-43.

Cohen L. J. (ed.), 1982, *Logic, Methodology, and Philosophy of Science VI: Proceedings of the Sixth International Congress of Logic, Methodology, and Philosophy of Science, Hannover, 1979*, Amsterdam\Warszawa: North-Holland\PWN.

Cohen L. J., and Hesse M. (eds.), 1980, *Applications of Inductive Logic: Proc 1978*, Oxford: Clarendon Press.

Cohen S., 1995, "Is There an Issue about Justified Belief?", *Philosophical Topics* 23, 113-127.

Constantini D., 1989, "Objectivism and Subjectivism in the Foundations of Statistics", *Erkenntnis* 31, 387-396.

Constantini D., and Galavotti M. C. (eds.), 1997, *Probability, Dynamics and Causality: Essays in Honour of Richard C. Jeffrey*, Dordrecht: Kluwer.

Cooke R. M., 1986, "Conceptual Fallacies in Subjective Probability", *Topoi* 5, 21-7.

Craig E. (ed.), 1998, *Routledge Encyclopedia of Philosophy*, London: Routledge.

Creath R., 1990, "Introduction", in Quine and Carnap 1990, 1-43.

Creath R., 1991, "Every Dogma Has Its Day", in Spohn (ed.) 1991a, 347-389.

Creath R., 1992, "Carnap's Conventionalism", in Sarkar (ed.) 1992a, 141-65.
Creath R., 1995/1996, "Are Dinosaurs Extinct?", *Foundations of Science* 1, 285-97.
Creath R., 1998, "Carnap, Rudolf (1891-1970)", in Craig (ed.), Vol. II, 208-15.
Crossley N. J., and Dummett M. A. E. (eds.), 1965, *Formal Systems and Recursive Functions: Proceedings of the Eighth Logic Colloquium, Oxford 1963*, Amsterdam: North–Holland.
Dancy J., and Sosa E. (eds.), 1993, *A Companion to Epistemology*, Oxford: Blackwell.
Davidson D., 1983, "Coherence Theory of Truth and Knowledge", in D. Henrich (ed.) 1983, *Kant oder Hegel?*, Stuttgart: Klett-Cotta, 423-38.
Descartes R., 1967, *The Philosophical Works of Descartes,* 2 vols. Translated by E.S. Haldane and G.R.T. Ross, Cambridge: Cambridge University Press.
De Finetti B., 1937/1964, "Foresight: Its Logical Laws, Its Subjective Sources". Translated from Italian and reprinted in Kyburg and Smokler (eds.) 1964b, 97-158.
De Finetti B., 1980, "On the Condition of Partial Exchangeability", in Jeffrey (ed.) 1980b, 193-205.
De Finetti B., 1982, "Probability: The Different Views and Terminologies in a Critical Analysis", in Cohen (ed.) 1982, 391-94.
De Finetti B., 1931/1989, "Probabilism". Translated from Italian and reprinted in *Erkenntnis* 31, 169-223.
Diaconis P., and Holmes S., 1997, "Are There Still Things to do in Bayesian Statistics", in Constantini and Galavotti (eds.) 1997, 5-18.
Dretske F. I., 1971a, "Conclusive Reasons", *Australasian Journal of Philosophy* 49, 1-22.
Dretske F. I., 1971b, "Reasons, Knowledge, and Probability", *Philosophy of Science* 38, 216-20.
Dubucs J.-P., 1993a, "Inductive Logic Revisited", in Dubucs (ed.) 1993b, 79-108.
Dubucs J.-P. (ed.), 1993b, *Philosophy of Probability*, Dordrecht: Kluwer.
Dumitriu A., 1977, *History of Logic*, Vol. III, Tunbridge Wells: Abacus.
Earman J. (ed.), 1983, *Testing Scientific Theories*, Minneapolis: University of Minnesota Press.
Earman J., 1992, *Bayes or Bust?: A Critical Examination of Bayesian Confirmation Theory*, Cambridge: MIT Press.
Earman J., and Salmon W. C., 1992, "On the Confirmation of Scientific Hypotheses", in M. H. Salmon, J. Earman, C. Glymour, J.G. Lennox, J.E. McGuire, J.D. Norton, W.C. Salmon, K.F. Schaffner (eds.) 1992, *Introduction to the Philosophy of Science*, Englewood Cliffs: Prentice Hall, 43-103.
Eells E., 1994, "Bayesian Epistemology: Probabilistic Confirmation and Rational Decision", *Protosoziologie* 6, 38-60.
Elgin C. Z. (ed.), 1997, *Nelson Goodman's New Riddle of Induction*, New York: Garland.
Emmerich D. S., and Greeno J. G., 1966, "Some Decision Factors in Scientific Investigation", *Philosophy of Science* 33, 262-70.
Essler W. K., 1970, *Induktive Logik. Grundlagen und Voraussetzungen*, Freiburg: Karl Alber.
Essler W. K., 1975, "Hintikka versus Carnap", in Hintikka (ed.) 1975b, 365-69.
Essler W. K., 1986, "How to Apply and Justify Inductive Logic", *Erkenntnis* 24, 47-55.
Feigl H., 1975, "Homage to Carnap", in Hintikka (ed.) 1975b, XII-XVII.
Feigl H., and Maxwell G. (eds.), 1961, *Current Issues in the Philosophy of Science. Proceedings of Section L of the American Association for the Advancement of Science 1959*, New York: Holt, Rinehart, Winston.
Feigl H., and Sellars W. (eds.), 1949, *Readings in Philosophical Analysis*, New York: Appleton–Century–Crofts.

Feigl H., Sellars W., and Lehrer K. (eds.), 1972, *New Readings in Philosophical Analysis*, New York: Meredith.

Feldman R., 1985, "Reliability and Justification", *The Monist* 68, 159-74.

Feldman R., 1993, "Evidence", in Dancy and Sosa (eds.) 1993, 119-122.

Feldman R., and Conee R., 1985, "Evidentialism", *Philosophical Studies* 48, 15-34.

Ferguson A., 1974, "Prior Distributions on Spaces of Probability Measures", *Annals of Statistics* 2, 615-29.

Festa R., 1993, *Optimum Inductive Methods: A Study In Inductive Probability, Bayesian Statistics, and Verisimilitude*, Dordrecht: Kluwer.

Festa R., 1995, "Verisimilitude, Disorder, and Optimum Prior Probabilities", in Kuipers and Mackor (eds.) 1995, 299-320.

Festa R., 1997, "Analogy and Exchangeability in Predictive Inferences", *Erkenntnis* 45, 229-252.

Festa R., 1999a, "Bayesian Confirmation", in M. C. Galavotti, and A. Pagnini (eds.) 1999, *Experience, Reality, and Scientific Explanation*, Dordrecht: Kluwer, 55-87.

Festa R., 1999b, "Scientific Values, Probability and Acceptance", in R. R. Favretti, G. Sandri, and R. Scazzieri (eds.) 1999, *Incommensurability and Translation: Kuhnian Perspectives on Scientific Communication and Theory Change*, Cheltenham: Elgar, 323-329.

Festa R., (forthcoming), "Induction, Probability, and Bayesian Epistemology", in L. Haaparanta, I. Niiniluoto (ed.), *Finnish Analytical Philosophy*, Amsterdam: Rodopi.

Fetzer J. H. (ed.), 1988, *Probability and Causality: Essays in Honor of Wesley C. Salmon*, Dordrecht: Reidel, Fetzer J. H. (ed.), 1993, *Foundations of Philosophy of Science: Recent Developments*, New York: Paragon House.

Fine T. L., 1973, *Theories of Probability: An Examination of Foundations*, New York: Academic Press.

Fine T. L., 1977, "An Argument for Comparative Probability", in Hintikka and Butts (eds.) 1977, 105-19.

Fisak J. E., and Pidgeon N., 1997, "The Conjunction Fallacy: The Case for the Existence of Competing Heuristic Strategies", *British Journal of Psychology* 88, 1-27.

Fitelson B., 1998, "The Plurality of Bayesian Measures of Confirmation and the Problem of Measure Sensitivity", *PSA 1998*.

Floistad G. (ed.), 1981, *Contemporary Philosophy: A New Survey*, Vol.1, *Philosophy of Language, Philosophical Logic*, The Hague: Nijhoff.

Fodor J., 1978/1981, "Propositional Attitudes", *Monist* 61. Reprinted in N. Block (ed.) 1981, *Readings in Philosophy of Psychology*, Vol. 2, London: Methuen, 45-63.

Fodor J., and Lepore E., 1992, *Holism: A Shopper's Guide*, Oxford: Blackwell.

Foley R., 1998, "Justification, Epistemic", in Craig (ed.) 1998, Vol. V, 157-65.

Franklin J., 2001, "Resurrecting logical probability", *Erkenntnis* 55, 277-305.

Friedman M., 1992, "Epistemology in the *Aufbau*", in Sarkar (ed.) 1992a, 15-57.

Fuller S., 1989, *Philosophy of Science and Its Discontents*, Boulder: Westview.

Gaifman H., Snir M., 1982, "Probabilities over Rich Languages", *Journal of Symbolic Logic* 47, 495-548.

Galavotti M. C., 1995/1996, "Operationism, Probability and Quantum Mechanics", *Foundations of Science* 1, 99-118.

Galavotti M. C., 1997, "Probabilism and Beyond", in Constantini and Galavotti (eds.) 1997, 113-25.

Gillies D. A., 1991, "Intersubjective Probability and Confirmation Theory", *British Journal for the Philosophy of Science* 42, 513-33.

Goldman A. H., 1988, *Empirical Knowledge*, Berkeley: University of California Press.

Goldman A. I., 1979/1993, "What is Justified Belief?", in G. Pappas (ed.) 1979, *Justification and Knowledge*, Dordrecht: Reidel 1979, 1-23. Reprinted in Goodman and Snyder (eds.) 1993, 89-104.

Goldman A. I., 1986, *Epistemology and Cognition*, Cambridge: Harvard University Press.

Goldman A. I., 1993, "Reliabilism", in Dancy and Sosa (eds.) 1993, 433-36.

Goldman A. I., 1998, "Reliabilism", in Craig (ed.) 1998, Vol. VIII, 204-208.

Goldstick D., 1989, "The Meaning of 'Grue'", *Erkenntnis* 31, 139-41.

Goodman M. F., and Snyder R. A. (eds.), 1993, *Contemporary Readings in Epistemology*, Englewood Cliffs: Prentince-Hall.

Goodman N., 1946, "A Query on Confirmation", *The Journal of Philosophy* 43, 383-385.

Goodman N., 1947/1997, "On Infirmities of Confirmation-Theory", *Philosophy and Phenomenological Research* 8, 149-151. Reprinted in Elgin (ed.) 1997, 17-19.

Goodman N., 1955/1983, *Fact, Fiction, and Forecast*. Fourth edition, Cambridge: Harvard University Press.

Grayling A. C., 1995, "Justification and the Structure of Inference", in *La Justification Épistémique* 1995, 1-19.

Grünbaum A., and Salmon W. C. (ed.), 1988 , *The Limitations of Deductivism*, Berkeley: University of California Press.

Hall N., 1994, "Correcting the Guide to Objective Chance", *Mind* 1-3, 505-17.

Haaparanta L., Kusch M., and I. Niiniluoto (eds.), 1990, *Language, Knowledge, and Intentionality: Perspectives on the Philosophy of Jaakko Hintikka*, Helsinki: Philosophical Society of Finland.

Hacking I., 1968, "One Problem about Induction", in Lakatos (ed.) 1968b, 44-59.

Hacking I., 1969, "Linguistically Invariant Inductive Logic", *Synthese* 20, 25-47.

Hacking I., 1970/1971, "The Leibniz-Carnap Program for Inductive Logic", *The Journal of Philosophy* 68, 597-610.

Hacking I., 1975, *The Emergence of Probability: A Philosophical Study of Early Ideas About Probability, Induction and Statistical Inference*, London: Cambridge University Press.

Hacking I., 1993/1997, "Goodman's New Riddle is Pre-Humean", *Revue Internationale de Philosophie* 47, 229-43. Reprinted in Elgin (ed.) 1997, 159-243.

Hailperin T., 1990, "Probability Logic in the Twentieth Century", *History and Philosophy of Logic*, 71-110.

Hajduk Z., 1984, *O akceptacji teorii empirycznej*, Lublin: RWKUL.

Harman G., 1993, "Induction: Enumerative and Hypothetical", in Dancy and Sosa (eds.) 1993, 201-206.

Hausman A., and Wilson F., 1967, *Carnap and Goodman: Two Formalists*, Iowa City: University of Iowa.

Heidelberger H., 1963, "Knowledge, Certainty, and Probability", *Inquiry* 6, 245–255.

Heil J., 1993, "Belief", in Dancy and Sosa (eds.) 1993, 45-48.

Hempel C. G., 1934, "On the Logical Positivist's Theory of Truth", *Analysis* 2, 49-59.

Hempel C. G., 1943, "A Purely Syntactical Definition of Confirmation", *The Journal of Symbolic Logic* 8, 122-43.

Hempel C. G., 1945/1949a, "Geometry and Empirical Science", *American Mathematical Monthly* 52. Reprinted in Feigl and Sellars (eds.) 1949, 238-249.

Hempel C. G., 1945/1949b, "On the Nature of Mathematical Truth", *American Mathematical Monthly* 52. Reprinted in Feigl and Sellars (eds.) 1949, 222-237.

Hempel C. G., 1945/1965, "Studies in the Logic of Confirmation", *Mind* 54. Reprinted in Hempel 1965, 3-46.

Hempel C. G., 1950/1996, "Problems and Changes in the Empiricist Criterion of Meaning", *Revue Internationale de Philosophie* 4, 41-63. Reprinted in Sahotra (ed.) 1996, 107-129.

Hempel C. G., 1959, "The Empiricist Criterion of Meaning", *Revue Internationale de Philosophie* 4. Reprinted in A. J. Ayer (ed.) 1959, 108-129.

Hempel C. G., 1961, "Deductive–Nomological versus Statistical Explanation", in Feigl and Maxwell (ed.) 1961, 98–169.

Hempel C. G., 1962, "Inductive Inconsistencies", in Kazemier and Vuysje (eds.) 1962, 128-58.

Hempel C. G., 1963, "Implications of Carnap's Work for the Philosophy of Science", in Schilpp (ed.) 1963, 685-709.

Hempel C. G., 1965, *Aspects of Scientific Explanation*, New York: The Free Press.

Hempel C. G., 1967, "Confirmation: Qualitative Aspects", in P. Edwards (ed.) 1967, *Encyclopedia of Philosophy*, New York: Macmillan, Vol. II, 185-87.

Hempel C. G., 1973, "Rudolf Carnap, Logical Empiricist", *Synthese* 25, 256-68.

Hempel C. G., 1975, "Homage to Carnap", in Hintikka (ed.) 1975b, XVIII-XXI.

Hesse M., 1964, "Analogy and Confirmation Theory", *Philosophy of Science* 31, 319-27.

Hesse M., 1968a, "Consilience of Inductions", in Lakatos (ed.) 1968b, 232-46.

Hesse M., 1968b, "Enumeration, Elimination and Analogy", in Lakatos (ed.) 1968b, 220-23.

Hesse M., 1969, "Positivism and the Logic of Scientific Theories", in Achinstein and Barker (eds.) 1969, 85-114.

Hesse M., 1980, "What is the Best Way to Assess Evidential Support", in Cohen and Hesse (eds.) 1980, 202-17.

Hetherington S. C., 1996, *Knowledge Puzzles: An Introduction to Epistemology*, Boulder: Westview.

Hilpinen R., 1966, "On Inductive Generalization in Monadic First-Order Logic with Identity", in Hintikka and Suppes (eds.) 1966, 133-154.

Hilpinen R., 1968, *Rules of Acceptance and Inductive Logic*, Amsterdam: North-Holland.

Hilpinen R., 1971, "Relational Hypotheses and Inductive Inference", *Synthese* 23, 266-286.

Hilpinen R., 1972, "Decision–Theoretic Approaches to Rules of Acceptance", in R. E. Olson, and A. M. Paul (eds.) 1972, *Contemporary Philosophy in Scandinavia*, Baltimore: The Johns Hopkins Press, 147–168.

Hilpinen R., 1973, "Carnap's New System of Inductive Logic", *Synthese* 25, 307-333.

Hilpinen R., 1977, "Remarks on Personal and Impersonal Knowledge", *Canadian Journal of Philosophy* 7, 1-9.

Hilpinen R., 1982, "Schlick on the Foundations of Knowledge", *Grazer Philosopische Studien* 16/17, 63-78.

Hilpinen R., 1999, "Hempel on the Problem of Induction", a paper presented at the Second Summer School for the Theory of Knowledge, M[1]dralin 1999.

Hilpinen R., and Hintikka J., 1971, "Rules of Acceptance, Indices of Lawlikeness, and Singular Inductive Inference: Reply to a Critical Discussion", *Philosophy of Science* 38, 303–307.

Hintikka J., 1953, *Distributive Normal Forms in the Calculus of Predicates*, Helsinki: Acta Philosophica Fennica 6.

Hintikka J., 1965a, "Distributive Normal Forms in First–Order Logic", in N.J. Crossley, and M.A.E. Dummett (eds.) 1965, 47–90.

Hintikka J., 1965b, "Towards a Theory of Inductive Generalization", in Bar–Hillel (ed.) 1965, 274–288.

Hintikka J., 1966, "A Two–Dimensional Continuum of Inductive Methods", in Hintikka and Suppes (eds.) 1966, 113–132.

Hintikka J., 1968a, "Conditionalization and Information", in Lakatos (ed.) 1968b, 303-306.

Hintikka J., 1968b, "Induction by Enumeration and Induction by Elimination", in Lakatos (ed.) 1968b, 191-216.

Hintikka J., 1968c, "Reply", in Lakatos (ed.) 1968b, 223-231.

Hintikka J., 1968d, "The Possibility of Rules of Acceptance", in Lakatos (ed.) 1968b, 144-46.

Hintikka J., 1969, "Inductive Independence and the Paradoxes of Confirmation", in Rescher et al. (eds.) 1969, 24–46.

Hintikka J., 1971, "Unknown probabilities, Bayesianism, and de Finetti's Representation Theorem", in Buck and Cohen (eds.) 1971, 325-41.

Hintikka J., 1973/1975, "Carnap's Heritage in Logical Semantics", *Synthese* 25, 372-97. Reprinted in Hintikka (ed.) 1975b, 217-42.

Hintikka J., 1975a, "Carnap and Essler versus Inductive Generalization", in Hintikka (ed.) 1975b, 371-80.

Hintikka J. (ed.), 1975b, *Rudolf Carnap, Logical Empiricist: Materials and Perspectives*, Dordrecht: Reidel.

Hintikka J., 1980, "Aristotelian Induction", *Revue Internationale de Philosophie* 133-34, 422-39.

Hintikka J., 1981, "Semantics: A Revolt Against Frege", in Floistad (ed.) 1981, 57-82.

Hintikka J., 1987a, "Replies and Comments", in Bogdan (ed.) 1987, 277-344.

Hintikka J., 1987b, "Self-Profile", in Bogdan (ed.) 1987, 3-40.

Hintikka J., 1991a, "Carnap, the Universality of Language and Extremality Axioms", in Spohn (ed.) 1991a, 325-336.

Hintikka J., 1991b "Kant's 'New Method of Thought' and His Theory of Mathematics", in J. Hintikka 1991, *Knowledge and the Known: Historical Perspectives in Epistemology*, Dordrecht: Reidel, 126-34.

Hintikka J., 1997, "Replies", in Sintonen (ed.) 1997b, 309-338.

Hintikka J., and Butts R. E. (eds.), 1977, *Proceedings of the Fifth International Congress of Logic, Methodology, and Philosophy of Science, London, Ontario, Canada, 1975*, Dordrecht: Reidel.

Hintikka J., and Hilpinen R., 1966, "Knowledge, Acceptance, and Inductive Logic", in Hintikka and Suppes (eds.) 1966, 1–20.

Hintikka J., and Niiniluoto I., 1976/1980, "An Axiomatic Foundation for the Logic of Inductive Generalization", in M. Przełęcki et al. (eds.) 1976, 57-81. A corrected version reprinted in Jeffrey (ed.) 1980b, 157-181.

Hintikka J., and Rantala V., 1975, "Systematizing Definability Theory", in S. Kanger (ed.) 1975, *Proceedings of the Third Scandinavian Logic Symposium*, Amsterdam: North-Holland, 42-62.

Hintikka J., and Suppes P. (eds.), 1966, *Aspects of Inductive Logic*, Amsterdam: North–Holland.

Holland J. H., Holyoak K. J., Nisbett R. E., and Thagard P. R., 1986, *Induction: Processes of Inference, Learning, and Discovery*, Cambridge: MIT Press.

Horwich P., 1993, "Bayesianism", in Dancy and Sosa (eds.) 1993, 40-44.

Hosack J., 1991, "On Probabilism and Induction", *Topoi* 227-228.

Hosiasson-Lindenbaum J., 1940, "On Confirmation", *The Journal of Symbolic Logic* 5, 133-48.

Howson C., 1990a, "Fitting Your Theory to the Facts: Probably Not Such a Bad Thing After All", in Savage (ed.) 1990, 224-44.

Howson C., 1990b, "Some Further Reflections on the Popper-Miller Disproof of Probabilistic Induction", *Australasian Journal of Philosophy* 68, 221-228.

Howson C., 1991a, "The Last Word on Induction?", *Erkenntnis* 73-82.
Howson C., 1991b, "The 'Old Evidence' Problem", *British Journal for the Philosophy of Science* 42, 547-555.
Howson C., 1993, "Personalistic Bayesianism", in Dubucs (ed.) 1993b, 1-12.
Howson C., 1995, "Theories of Probability", *British Journal for the Philosophy of Science* 46, 1-32.
Howson C., 1997a, "A Logic of Induction", *Philosophy of Science* 64, 268-290.
Howson C., 1997b, "Logic and Probability", *British Journal for the Philosophy of Science* 48, 517-31.
Howson C., and Urbach P., 1993, *Scientific Reasoning: The Bayesian Approach*, La Salle: Open Court.
Humburg J., 1986, "The Solution of Hempel's Raven Paradox in Rudolf Carnap's System of Inductive Logic", *Erkenntnis* 24, 57-72.
Hume D.,1739/1967, *A Treatise of Human Nature*, Oxford: Clarendon.
Hume D., 1748/1963, *An Enquiry Concerning Human Understanding*, La Salle: Open Court.
Humphreys P., 1995/1996, "Computational Empiricism", *Foundations of Science* 1, 119-30.
Humphreys P., 1998, "Probability, Interpretations of", in Craig (ed.) 1998, Vol. VII, 701-705.
Ingarden R., 1962, "Bemerkungen zum Problem der Begründung", *Studia Logica* 13, 153-72.
Jeffrey R. C., 1956, "Valuation and Acceptance of Scientific Hypotheses", *Philosophy of Science* 23, 237–246.
Jeffrey R. C., 1957, "Contributions to the Theory of Inductive Probability", Princeton: PhD Thesis.
Jeffrey R., 1965, *The Logic of Decision*, New York: McGraw-Hill.
Jeffrey R. C., 1966/1997, "Goodman's Query", *The Journal of Philosophy* 63, 281-88. Reprinted in Elgin (ed.) 1997, 79-86.
Jeffrey R. C., 1968a, "Probable Knowledge", in Lakatos (ed.) 1968b, 166-80.
Jeffrey R. C., 1968b, "Reply", in Lakatos (ed.) 1968b, 189-90.
Jeffrey R. C., 1973, "Carnap's Inductive Logic", *Synthese* 25, 299-306.
Jeffrey R. C., 1975, "Homage to Carnap", in Hintikka (ed.) 1975b, XXI-XXIV.
Jeffrey R. C., 1977, "Misses Redux", in Hintikka and Butts (eds.) 1977, 213-22.
Jeffrey R. C., 1980a, "Introduction", in Jeffrey (ed.) 1980b, 1-6.
Jeffrey R. C. (ed.), 1980b, *Studies in Inductive Logic and Probability,* Vol. 2, Berkeley: University of California Press.
Jeffrey R. C., 1981, "Choice, Chance, and Credence", in Floistad (ed.) 1981, 367-86.
Jeffrey R. C., 1983, "Bayesianism with a Human Face", in J. Earman (ed.) 1982, 133-56.
Jeffrey R. C., 1984, "The Impossibility of Inductive Probability", *Nature* 310, 433.
Jeffrey R. C., 1985/1992, "Probability and the Art of Judgment", in P. Achinstein, and M. Hannaway (eds.) 1985, *Observation and Experiment in Modern Science*, Cambridge: MIT Press, 133-56. Reprinted in (Jeffrey 1992), 44-76.
Jeffrey R. C., 1986a, "Judgmental Probability and Objective Chance", *Erkenntnis* 24, 5-16.
Jeffrey R. C., 1986b, "Probabilism and Induction", *Topoi* 5, 51-58.
Jeffrey R. C., 1987/1992, "Alias Smith and Jones: The Testimony of the Senses", *Erkenntnis* 26, 391-399. Reprinted in Jeffrey 1992, 108-16.
Jeffrey R. C., 1989, "Reading "Probabilismo"", *Erkenntnis* 31, 225-237.
Jeffrey R. C., 1991, "After Carnap", in Spohn (ed.) 1991a, 255-262.
Jeffrey R. C., 1991/1992, "Radical Probabilism", in Jeffrey 1992, 1-13.
Jeffrey R. C., 1992, *Probability and the Art of Judgment*, New York: Cambridge University Press.

Jeffrey R. C., 1995, "Probability Reparation: The Problem of New Explanation", *Philosophical Studies* 77, 97-101.

Jeffrey R. C., 1997, "Unknown Probabilities: In Memory of Annemarie Anrod Shimony (1928-1995)", *Erkenntnis* 45, 327-335.

Jeffrey R. C., and Carnap R., 1957/1958, "Correspondence", Manuscript.

Jeffreys H., 1939, *The Theory of Probability*, Oxford: Oxford University Press.

Johnson W. E., 1924, *Logic. Part III*, Cambridge.

Johnson W. E., 1932, "Appendix to Probability: The Deductive and Inductive Problems", *Mind* 41, 409-23.

Kahane H., 1965, "Nelson Goodman's Entrenchment Theory", *Philosophy of Science* 32, 377-83.

Kahneman D., Slovic P., and Tversky A. (eds.), 1982, *Judgments under Uncertainty: Heuristics and Biases*, Cambridge: Cambridge University Press.

Kakkuri-Knuuttila M.-L., 1997, "What Can the Sciences of Man Learn from Aristotle?", in M. Sintonen (ed.) 1997b, 19-39.

Kakkuri-Knuuttila M.-L., and Knuuttila S., 1990, "Induction and Conceptual Analysis in Aristotle", in Haaparanta et al. (eds.) 1990, 294-303.

Kaplan M., 1996, *Decision Theory as Philosophy*, Cambridge: Cambridge University Press.

Kaplan M., 1998, "Induction, Epistemic Issues in", in Craig (ed.) 1998, Vol. IV, 745-52.

Kawalec P., 1997a, "Two Concepts of Epistemic Justification", in R. Wójcicki (ed.) 1997, 85-96.

Kawalec P., 1997b, "Pragmatic Predicaments to the Logical Theory of Nondeductive Inferential Justification", in G. Schurz and G. Dorn (eds.) 1997, *Preproceedings of the 20th International Wittgenstein Symposium*, Kirchberg 1997, pp. 452-457.

Kawalec P., 1997c, "Teza Laudana a internalizm Jaakko Hintikki teorii indukcji", *Roczniki Filozoficzne* 65, *f.* 1, 75-87.

Kawalec P., 1998, "A Pragmatic Flaw in Inferential Epistemic Justification", *Reports on Philosophy* 18(1998), 71-81.

Kawalec P., 1999, "The Rise of the Concept of Justification in the Lvov-Warsaw School and the Revision of Psychologism in Logic (I)", in J. Nida-Rümelin (ed.) 1999, *Rationality, Realism, Revision*, New York: de Gruyter 1999, 862-72.

Kawalec P., 2000a, "Cognitive Merit and Axiomatic Foundations of Inductive Probability: A Trade-off?", in V. Fano, G. Tarozzi, M. Stanzione (eds.), *Prospettive della logica e della filosofia della scienza*, Rubbettino: Pubblicazioni della SILFS, 45-56.

Kawalec P., 2000b, "Ontologiczna interpretacja pojęcia prawdopodobieństwa", *Studia Metafilozoficzne* 2(2000), forthcoming.

Kazemier B. H. and Vuysje D. (eds.), 1962, *Logic and Language: Studies Dedicated to Professor Rudolf Carnap on the Occasion of His Seventieth Birthday*, Dordrecht: Reidel.

Kemeny J. G., 1951, "Carnap on Probability", *Review of Metaphysics* 5, 145-56.

Kemeny J. G., 1952, "Extension of the Methods of Inductive Logic", *Philosophical Studies* 3, 38-42.

Kemeny J. G., 1953a, "A Contribution to Inductive Logic", *Philosophy and Phenomenological Research* 13, 371-374.

Kemeny J. G., 1953b, "Review of R. Carnap, The Continuum of Inductive Methods", *The Journal of Symbolic Logic* 18, 168-69.

Kemeny J. G., 1953c, "The Logical Measure Function", *The Journal of Symbolic Logic* 18, 289-308.

Kemeny J. G., 1953d, "The Use of Simplicity in Induction", *Philosophical Review* 62, 391–408.

Kemeny J. G., 1955a, "Fair Bets and Inductive Probabilities", *The Journal of Symbolic Logic* 20, 263-73.
Kemeny J. G., 1955b, "Two Measures of Complexity", *The Journal of Philosophy* 52, 722-33.
Kemeny J. G., 1963, "Carnap's Theory of Probability and Induction", in P. A. Schilpp (ed.), The Philosophy of Rudolf Carnap, La Salle: Open Court, 711-38.
Keynes J. M, 1921/1948, *A Treatise on Probability*, London: MacMillan.
Kim J., 1988, "What is 'Naturalized Epistemology'?", *Philosophical Perspectives* 2, 381-405.
Kistler M., 1995, "La justification de la connaissance empirique dans le Cercle de Vienne", in *La Justification Épistémique* 1995, 1-28.
Kitcher P., 1992, "The Naturalists Return", *The Philosophical Review* 101, 53–114.
Klayman J., and Ha Y.-W., 1987, "Confirmation, Disconfirmation, and Information in Hypothesis Testing", *Psychological Review* 94, 211-28.
Koj L., 1987, "Pragmatyka logiczna", in W. Marciszewski (ed.) 1987, *Logika formalna*, 281-95.
Koj L., 1990, *Problemy semiotyki logicznej*, Warszawa: PTS.
Kolmogorov A. N., 1946 , *Foundations of the Theory of Probability*, New York: Chelsea.
Koopman N., 1940, "The Axioms and Algebra of Intuitive Probability", *Annals of Mathematics* 41, 269-92.
Koppelberg D., 1998, "Foundationalism and Coherentism Reconsidered", *Erkenntnis* 49, 255-83.
Kornblith H., 1993, *Inductive Inference and Its Natural Ground: An Essay in Naturalistic Epistemology*, Cambridge: MIT Press.
Kronz F. M., 1992, "Carnap and Achinstein on Evidence", *Philosophical Studies* 67, 151-167.
Kuhn T., 1971, "Notes on Lakatos", in Buck and Cohen (eds.) 1971, 137-46.
Kuipers T. A. F., 1973/1975, "A Generalization of Carnap's Inductive Logic", *Synthese* 25, 334-336. Reprinted in Hintikka (ed.) 1975b, 361-63.
Kuipers T. A. F., 1976a, "A Two-dimensional Continuum of A Priori Probability Distributions on Constituents", in Przełęcki et al. (eds.) 1976, 82-92.
Kuipers T. A. F., 1976b, "Inductive Probability and the Paradox of Ideal Evidence", *Philosophica* 17, 197-205 .
Kuipers T. A. F., 1978a, "On the Generalization of the Continuum of Inductive Methods to Universal Hypotheses.", *Synthese* 37, 255-284.
Kuipers T. A. F., 1978b, *Studies in Inductive Probability and Rational Expectation*, Dordrecht: Reidel 1978.
Kuipers T. A. F., 1980, "A Survey of Inductive Systems", in Jeffrey (ed.) 1980b, 183-192.
Kuipers T. A. F., 1983, "Non-inductive Explication of Two Inductive Intuitions", *British Journal for the Philosophy of Science* 34, 209-223.
Kuipers T. A. F., 1984a, "An Approximation of Carnap's Optimum Estimation Method", *Synthese* 61, 361-362.
Kuipers T. A. F., 1984b, "Two Types of Inductive Analogy By Similarity", *Erkenntnis* 21, 63-88.
Kuipers T. A. F., 1986, "Some Estimates of the Optimum Inductive Method", *Erkenntnis* 24, 37-46.
Kuipers T. A. F., 1988, "Inductive Analogy By Similarity and Proximity", in D. H. Helman (ed.) 1988, *Analogical Reasoning*, Dordrecht: Kluwer, 299-313.
Kuipers T. A. F., and Mackor A. R. (eds.), 1995, *Cognitive Patterns in Science and Common Sense*, Rodopi: Amsterdam.

Kuipers T. A. F., 1997, "The Carnap-Hintikka Program in Inductive Logic", in Sintonen (ed.) 1997b, 87-99.

Kuipers T.A.F., 1998, "Confirmation Theory", in W. Craig (ed.) 1998, Vol. II, 532-36.

Kuipers T. A. F., 2000, *From Instrumentalism to Constructive Realism*, Dordrecht: Kluwer.

Kusch M., 1995, *Psychologism: A Case Study in the Sociology of Philosophical Knowledge*, London: Routledge.

Kyburg H. E., 1960/1961, "Demonstrative Induction", *Philosophy and Phenomenological Research* 21, 80-92.

Kyburg H. E., 1961, *Probability and the Logic of Rational Belief*, Middletown: Wesleyan University Press.

Kyburg H. E., 1964, "Recent Work in Inductive Logic", *American Philosophical Quarterly* 1, 249-287.

Kyburg H. E., 1965, "Probability, Rationality, and a Rule of Detachment", in Bar-Hillel (ed.) 1965, 301–310.

Kyburg H. E., 1966, "Probability and Decision", *Philosophy of Science* 33, 250-61.

Kyburg H. E., 1967, "Bets and Beliefs", *The Journal of Philosophy* 64, 742-43.

Kyburg H. E., 1968a, "Reply", in Lakatos (ed.) 1968b, 161-65.

Kyburg H. E., 1968b, "The Rule of Detachment in Inductive Logic", in Lakatos (ed.) 1968b, 98-119.

Kyburg H. E., 1970, "Conjunctivists", in Swain (ed.) 1970, 55-82.

Kyburg H. E., 1976, "Local and Global Induction", in Bogdan (ed.) 1976, 191-215.

Kyburg H. E., 1980, "Inductive Logic and Experimental Design", in Cohen and Hesse (eds.) 1980, 90-101.

Kyburg H., 1987, "Hobgoblin", *Monist* 70, 141–151.

Kyburg H. E., and Smokler H. E. (eds.), 1964, *Studies in Subjective Probability*, New York: Wiley.

La Justification Épistémique, 1995, *Colloque International*, Université de Neuchatel: Séminaire de philosophie.

Ladyman J., 1998, "What is structural realism?", *Studies in History and Philosophy of Science* 29, 409-24.

Lakatos I., 1968a, "Changes in the Problem of Inductive Logic", in Lakatos (ed.) 1968b, 315-417.

Lakatos I. (ed.), 1968b, *The Problem of Inductive Logic. Proceedings of the International Colloquium in the Philosophy of Science, London, 1965,* Vol. 2, Amsterdam: North–Holland.

Lakatos I. (ed.) 1971, "History of Science and Its Rational Reconstructions", in Buck and Cohen (eds.) 1971, 91-136.

Lehrer K., 1970, "Justification, Explanation, and Induction", in Swain (ed.) 1970, 100-133.

Lehrer K., 1974, *Knowledge*, Oxford: Clarendon Press.

Lehrer K., 1980, "Truth, Evidence, and Error", in Cohen and Hesse (eds.) 1980, 130-142.

Lehrer K., 1981a, "A Self Profile", in Bogdan (ed.) 1981, 1-104.

Lehrer K., 1981b, "Replies", in Bogdan (ed.) 1981, 223-241.

Lehrer K., 1990, *Theory of Knowledge*, Boulder: Westview.

Lenz J. W., 1956, "Carnap on Defining "Degree of Confirmation"", *Philosophy of Science* 23, 230-36.

Levi I, 1976, "Acceptance Revisited", in Bogdan (ed.) 1976, 1-71.

Levi I., 1980, "Potential Surprise: Its Role in Inference and Decision Making", in Cohen and Hesse (eds.) 1980, 1-27.

Lewis C. I., 1946, *An Analysis of Knowledge and Valuation*, La Salle: Open Court.

Lewis D., 1971, "Immodest Inductive Methods", *Philosophy of Science* 38, 54-63.
Lewis D., 1980, "A Subjectivist's Guide to Objective Chance", in Jeffrey (ed.) 1980b, 263-293.
Lewis D., 1994/1999, "Humean Supervenience Debugged", in Lewis D., *Papers in Metaphysics and Epistemology*, Cambridge: Cambridge University Press, 224-47.
Loewer B., 1998, "Probability Theory and Epistemology", in Craig (ed.), Vol. VI, 705-11.
Luper S., 1998, "Belief and Knowledge", in Craig (ed.) 1998, Vol. I, 706-709.
Lycan W., 1998, "Theoretical (Epistemic) Virtues", in Craig (ed.) 1998, Vol. IX, 340-43.
Łakoma E., 1992, *Historyczny rozwój pojęcia prawdopodobieństwa*, Warszawa: SNM.
Łoś J., 1960, "O ciałach zdarzeń i ich definicji w aksjomatyce teorii prawdopodobieństwa", *Studia Logica* 9, 95-115.
Łukasiewicz J., 1913/1970, *Die logische Grundlagen der Wahrscheinlichkeitsrechnung*, Kraków 1913. Translated by O. Wojtasiewicz as "Logical Foundations of Probability Theory", in *Selected Works*, Amsterdam/Warszawa: North-Holland/PWN, 16-64. Edited by L. Borkowski.
Maher P., 1996, "Subjective and Objective Confirmation", *Philosophy of Science* 63, 149-74.
Maher P., 1997, "Depragmatized Dutch Book Arguments", *Philosophy of Science* 64, 291-305.
Maher P., 1998, "Inductive Inference", in Craig (ed.) 1998, Vol. IV, 755-59.
Maher P., 1999, "Inductive Logic and the Ravens Paradox", *Philosophy of Science* 66, 50-70.
Mann W., Peters A., and Thöle B., 1992, "Rechtfertigung", in J. Ritter and K. Gründer (eds.) 1992, *Philosophisches Wörterbuch*, Basel: Schwabe, Vol. VIII, 251-64.
Marciszewski W., 1972, *Podstawy logicznej teorii przekonań*, Warszawa: PWN.
Martin R. M., 1958, "A Formalization of Inductive Logic", *The Journal of Symbolic Logic* 23, 251-256.
Martin R. M., 1959, *Toward a Systematic Pragmatics*, Amsterdam: North–Holland.
Mazurkiewicz S., 1932, "Zur Axiomatik der Wahrscheinlichkeitsrechnung", *Comptes Rendus des Science de la Société des Sciences et des Lettres de Varsovie*, Vol. 25, 3, 1-4.
Michalos A. C., 1971a, "Hilpinen's Rules of Acceptance and Inductive Logic", *Philosophy of Science* 38, 293–302.
Michalos A. C., 1971b, *The Popper-Carnap Controversy*, The Hague: Nijhoff.
Miller D., 1968, "On the Subjective Interpretation of Probability", in Lakatos (ed.) 1968b, 69-72.
Miller D., 1990, "A Restoration of Popperian Inductive Skepticism", *British Journal for the Philosophy of Science* 57, 137-139.
Milton J. R., 1987, "Induction before Hume", *British Journal for the Philosophy of Science* 38, 49-74.
Misak C. J., 1996, *Verificationism: Its History and Prospects*, London: Routledge.
Mises R. von, 1957, *Probability, Statistics, and Truth*, New York: Dover.
Mondadori M., 1987, "Hintikka's Inductive Logic", in Bogdan (ed.) 1987, 157-180.
Mortimer H., 1982, *Logika indukcji. Wybrane problemy*, Warszawa: PWN.
Moulines C. U., 1991, "Making sense of Carnap's *Aufbau*", in Spohn (ed.) 1991a, 263-286.
Mura A., 1990, "When Probabilistic Support Is Inductive", *Philosophy of Science* 57, 278-289.
Musgrave A., 1989, "Deductivism versus Psychologism", in M. A. Noturno (ed.) 1989, *Perspectives on Psychologism,* Leiden: Brill, 315-340.
Nagel E., 1944/1949, "Logic Without Ontology", in Y. H. Krikorian (ed.), Naturalism and the Human Spirit, Columbia University Press. Reprinted in Feigl and Sellars (eds.) 1949, 191–210.

Nagel E., 1945, "Probability and Non-demonstrative Inference", *Philosophy and Phenomenological Research* 5, 485-507.
Nagel E., 1963, "Carnap's Theory of Induction", in Schilpp (ed.) 1963, 785-825.
Nagel E., Suppes P., and Tarski A., (eds.), 1962, *Logic, Methodology, and Philosophy of Science*, Stanford: Stanford University Press.
Nekrasas E., 1987/1992, *Wiedza prawdopodobna: Powstanie i rozwój w empiryzmie logicznym programu probabilistycznej oceny wiedzy naukowej*. Translated by Z. Simbierowicz, Warszawa: PWN.
Neurath O., 1955, "Principles of the Theory of Probability", in Neurath O., Carnap R., and Morris Ch. (eds.) 1955, *International Encyclopedia of Unified Science,* Chicago: Chicago University Press, Vol. II, 341-422.
Newton-Smith W. H., 1981, The Rationality of Science, London: Routledge.
Niiniluoto I., 1972, "Inductive Systematization: Definition and A Critical Survey", *Synthese* 25, 25-81.
Niiniluoto I., 1973, "Review of A. C. Michalos, *The Popper-Carnap Controversy*, The Hague: Nijhoff 1971", *Synthese* 25, 417-36.
Niiniluoto I., 1976, "Inductive Logic and Theoretical Concepts", in Przełęcki et al. (eds.) 1976, 93-112.
Niiniluoto I., 1977, "On a K-Dimensional System of Inductive Logic", in F. Suppe, and P. D. Asquith (eds.) 1977, *PSA 1976*, Vol. 2, 425-47.
Niiniluoto I., 1980, "Analogy, Transitivity/the Confirmation of Theories", in Cohen and Hesse (eds.) 1980, 218-234.
Niiniluoto I., 1981a, "Analogy and Inductive Logic", *Erkenntnis* 16, 1-34.
Niiniluoto I., 1981b, "Statistical Explanation Reconsidered", *Synthese* 48, 437-72.
Niiniluoto I., 1987, *Truthlikeness*, Dordrecht: Kluwer.
Niiniluoto I., 1988a, "Analogy and Similarity in Scientific Reasoning", in D. H. Helman (ed.) 1988, *Analogical Reasoning*, Dordrecht: Kluwer 271-298.
Niiniluoto I., 1988b, "Probability, Possibility, and Plenitude", in Fetzer (ed.) 1988, 91-108.
Niiniluoto I., 1994/1995, "Hintikka and Whewell on Aristotelian Induction", *Grazer Philosophische Studien* 49, 49-61.
Niiniluoto I., 1997, "Inductive Logic, Atomism, and Observational Error", in Sintonen (ed.) 1997b, 117-31.
Niiniluoto I., and Toumela R., 1973, *Theoretical Concepts and Hypothetico-Inductive Inference*, Dordrecht: Reidel.
Nozick R., 1981, *Philosophical Explanations*, Cambridge: Harvard University Press.
Oberdan T., 1998, "The Vienna Circle's 'Anti-Foundationalism'", *British Journal for the Philosophy of Science* 49, 297-308.
O'Donnell R., 1990, "The Epistemology of J. M. Keynes", *British Journal for the Philosophy of Science* 41, 333-50.
Pastin M., 1981, "Social and Anti-Social Justification: A Study of Lehrer's Epistemology", in Bogdan (ed.) 1981, 205-22.
Picardi E., 1997, "Sigwart, Husserl and Frege on Truth and Logic, or Is Psychologism Still a Threat?", *European Journal of Philosophy* 5, 162-182.
Pietarinen J., 1970, "Quantitative Tools for Evaluating Scientific Systematizations", in J. Hintikka, and Suppes P. (eds.) 1970, Information and Inference, Dordrecht: Reidel, 123-147.
Pietarinen J., 1972, *Lawlikeness, Analogy, and Inductive Logic*, Amsterdam: North-Holland.
Pietarinen J., 1974, "Inductive Immodesty and Lawlikeness", *Philosophy of Science* 41, 196-198.

Plantinga A., 1993, *Warrant: The Current Debate*, Oxford: Oxford University Press.
Plantinga A., 1995a, "Justification in the 20th Century", *Philosophy and Phenomenological Research, Suppl. Vol.* 50, 45-71.
Plantinga A., 1995b, "Précis of Warrant: The Current Debate and Warrant and Proper Function", *Philosophy and Phenomenological Research* 55, 393-464.
Plantinga A., 1995c, "Reliabilism, Analyses, and Defeaters", *Philosophy and Phenomenological Research* 55, 427-64.
Pollock J., 1973, "Laying the Raven to Rest: A Discussion of Hempel and the Paradoxes of Confirmation", *The Journal of Philosophy* 70, 747-54.
Pollock J., 1986, *Contemporary Theories of Knowledge*, Totowa: Rowman & Littlefield.
Popper K., 1968, "On Rules of Detachment and So-called Inductive Logic", in Lakatos (ed.) 1968b, 130-39.
Popper K., and Miller D., 1983, "A Proof of the Impossibility of Inductive Probability", *Nature* 302, 687-688.
Przełęcki M., and Wójcicki R., 1969/1977, "The Problem of Analyticity", *Synthese* 19. Reprinted in Przełęcki and Wójcicki (eds.) 1977, 589-614.
Przełęcki M., and Wójcicki R. (eds.), 1977, *Twenty-five Years of Logical Methodology in Poland*, Warszawa\Dordrecht: PWN\Reidel.
Przełęcki M., Szaniawski K., and Wójcicki R. (eds.), 1976, *Formal Methods in the Methodology of Empirical Sciences*, Wrocław/Dordrecht: Ossolineum/Reidel.
Psillos S., 1996, "Scientific Realism and the 'Pessimistic Induction'", *Philosophy of Science (Suppl.)* 63, S306-S314.
Putnam H., 1963, "Degree of Confirmation and Inductive Logic", in Schilpp (ed.) 1963, 761-783.
Quine W. V., 1950/1996, "Two Dogmas of Empiricism", *Philosophical Review* 60, 20-43. Reprinted in Sahotra (ed.) 1996, 18-41.
Quine W. V., 1962, "Carnap and Logical Truth", in Kazemier and Vuysje (eds.) 1962, 39-63.
Quine W. V., 1969, "Epistemology Naturalized", in Quine W. V. 1969, *Ontological Relativity and Other Essays*, New York: Columbia University Press, 69-90.
Quine W. V., 1975, "Homage to Carnap", in Hintikka (ed.) 1975b, XXIV-XXVII.
Quine W. V., and Carnap R., 1990, *Dear Carnap, Dear Van: The Quine–Carnap Correspondence and Related Work*. Edited by R. Creath, Berkeley: University of California Press.
Quine W. V., 1981, "Reply to Stroud", in P. A. French, T. E. Uehling, and H. K. Wettstein (eds.) 1981, *Midwest Studies in Philosophy VI*, Minneapolis: University of Minnesota, 473-75.
Ramsey F. P., 1931/1964, "Truth and Probability", in Ramsey F. P., 1931, *The Foundations of Mathematics and Other Logical Essays*. Edited by R. B. Braithwaite, London: Harcourt Brace. Reprinted in Kyburg and Smokler (eds.) 1964b, 61-92.
Ramsey F. P., 1989, "Mr Keynes on Probability". Edited by D. H. Mellor, *British Journal for the Philosophy of Science* 40, 219-222.
Rantala V., 1987, "Constituents", in Bogdan (ed.) 1987, 43-76.
Redhead M., 1985, "On the Impossibility of Inductive Probability", *British Journal for the Philosophy of Science* 36, 185-191.
Reichenbach H., 1935/1949, *The Theory of Probability: An Inquiry into the Logical and Mathematical Foundations of the Calculus of Probability*. Second edition translated by E. H. Hutten, and M. Reichenbach, Berkeley: University of California Press.
Reichenbach H., 1945, "Reply to Donald C. Williams' Criticism of the Frequency Theory of Probability", *Philosophy and Phenomenological Research* 5, 508-512

Rescher N., Anderson A.R., Benacerraf P., Grünbaum A., Massey G. J., and Rudner R.S. (eds.), 1969, *Essays in Honor of Carl G. Hempel: A Tribute on the Occasion of His Sixty-Fifth Birthday*, Dordrecht: Reidel.

Richardson A., 1992, "Logical Idealism and Carnap's Construction of the World", in Sarkar (ed.) 1992a, 59-92.

Richardson A., 1998, *Carnap's Construction of the World: The "Aufbau" and the Emergence of Logical Empiricism*, Cambridge: Cambridge University Press.

Ring E., 1986, "Goldman's Reliability Theory of Justified Belief", *Australasian Journal of Philosophy* 64, 501-507.

Rodriguez A. R., 1987, "On Popper-Miller's Proof of the Impossibility of Inductive Probability", *Erkenntnis* 27, 353-357.

Roorda J., 1997, "Fallibilism, Ambivalence, and Belief", *The Journal of Philosophy* 94, 126-55.

Rorty R., 1979, "The Unnaturalness of Epistemology", in D. F. Gustafson and B. Tapscott (eds.) 1979, *Body, Mind and Method*, Dordrecht: Reidel, 77-92.

Rosenkrantz R. D., 1980, "Rational Information Acquisition", in Cohen and Hesse (eds.) 1980, 68-89.

Rosenkrantz R. D., 1981, "Foundations and Applications of Inductive Probability", Atascadero: Ridgeview.

Rynin M., 1947, "Probability and Meaning", *The Journal of Philosophy* 44, 589-97.

Saarinen E., Hilpinen R., Niiniluoto I., and Hintikka M. P. (eds.), 1979, *Essays in Honour of Jaakko Hintikka: On the Occasion of His Fiftieth Birthday on January 12, 1979,* Dordrecht: Reidel.

Sahlin N.-E., 1997, "'He is No Good for My Work': On the Philosophical Relations between Ramsey and Wittgenstein", in Sintonen (ed.) 1997b, 61-84.

Salmon W. C., 1961, "Vindication of Induction", in Feigl and Maxwell (eds.) 1961, 245-256.

Salmon W. C., 1963a, "Inductive Inference", in B. H. Baumrin (ed.) 1963, *Philosophy of Science: The Delaware Seminar*, Vol. 2, New York: Wiley, 341-70.

Salmon W. C., 1963b, "On Vindicating Induction", *Philosophy of Science* 30, 252-61.

Salmon W. C., 1965a, "Consistency, Transitivity, and Inductive Support", *Ratio* 7, 164-69.

Salmon W. C., 1965b, "The Status of Prior Probabilities in Statistical Explanation", *Philosophy of Science* 32, 137-46.

Salmon W. C., 1966, "Use, Mention, and Linguistic Invariance", *Philosophical Studies* 17, 13-18.

Salmon W. C., 1967a, "Carnap's Inductive Logic", *The Journal of Philosophy* 64, 725-739.

Salmon W. C., 1967b, *The Foundations of Scientific Inference*, Pittsburgh: University of Pittsburgh Press.

Salmon W. C., 1968a, "Reply", in Lakatos (ed.) 1968b, 74-97.

Salmon W. C., 1968b, "The Justification of Inductive Rules of Inference", in Lakatos (ed.) 1968b, 24-43.

Salmon W. C., 1968c, "Who Needs Inductive Acceptance Rules?", in Lakatos (ed.) 1968b, 139-144.

Salmon W. C., 1969, "Partial Entailment as a Basis for Inductive Logic", in Rescher et al. (eds.) 1969, 47-82.

Salmon W. C., 1970, "Bayes's Theorem and the History of Science", in R. H. Stuewer (ed.) 1970, *Minnesota Studies in the Philosophy of Science,* Vol. V, Minneapolis: Minnesota University Press, 68-86.

Salmon W. C., 1974, "Russell on Scientific Inference or Will the Real Deductivist Please Stand Up?", in G. Nakhnikian (ed.) 1974, *Bertrand Russell's Philosophy*, London: Duckworth, 183-208.
Salmon W. C., 1975, "Confirmation and Relevance", in G. Maxwell, and R. M. Anderson (eds.) 1975, *Minnesota Studies in the Philosophy of Science,* vol. VI, Minneapolis: University of Minnesota Press, 3-36.
Salmon W. C., 1979, "Propensities: A Discussion Review", *Erkenntnis* 14, 183-216.
Salmon W. C., 1981, "In Praise of Relevance", *Teaching Philosophy* 4, 261-75.
Salmon W. C., 1988, "Dynamic Rationality: Propensity, Probability, and Credence", in Fetzer (ed.) 1987, 3-40.
Salmon W. C., 1989, "Four Decades of Scientific Explanation", in P. Kitcher, and Salmon W. C. (eds.) 1989, *Minnesota Studies in the Philosophy of Science*, Vol. XIII, *Scientific Explanation*, Minneapolis: University of Minnesota Press, 3-219.
Salmon W. C., 1990a, *Four Decades of Scientific Explanation*, Minneapolis: University of Minnesota Press.
Salmon W. C., 1990b, "Rationality and Objectivity in Science, or Tom Kuhn Meets Tom Bayes", in Savage (ed.) 1990, 175-204.
Salmon W. C., 1993, "Carnap Rudolf", in Dancy and Sosa (eds.) 1993, 56.
Sarkar S. (ed.), 1992a, *Carnap: A Centenary Reappraisal*, Dordrecht: Kluwer.
Sarkar S., 1992b, "Rudolf Carnap, 1891-1970: The Editor's Introduction", in Sarkar (ed.) 1992a, 1-14.
Sarkar S. (ed.), 1996, *Decline and Obsolescence of Logical Empiricism: Carnap vs. Quine and the Critics*, New York: Garland.
Savage L. J., 1967, "Implications of Personal Probability for Induction", *The Journal of Philosophy* 64, 593-607.
Savage W. C. (ed.) 1990, *Scientific Theories, Minnesota Studies in the Philosophy of Science*, Vol. XIV, Minneapolis: University of Minnesota Press.
Scheffler I., 1963/1967, *The Anatomy of Inquiry*. Second edition, New York: Alfred Knopf.
Schick F., 1967, "Confirmation: Quantitative Aspects", in P. Edwards (ed.) 1967, *Encyclopedia of Philosophy*, Vol. II, New York: Macmillan, 187-89.
Schick F., 1970, "Three Logics of Belief", in Swain (ed.) 1970, 6-26.
Schlick M., 1934/1959, "Über das Fundament der Erkenntnis", in *Erkenntnis* 4. Translated as "The Foundation of Knowledge" and reprinted in Ayer (ed.) 1959, 209-27.
Schlick M., 1935/1949, "On the Relation Between Psychological and Physical Concepts", *Revue de Synthese*. Translated by W. Sellars and reprinted in Feigl and Sellars (eds.) 1949, 393-407.
Schilpp P. A. (ed.), 1963, *The Philosophy of Rudolf Carnap*, La Salle: Open Court.
Scott D., and Krauss P., 1966, "Assigning Probabilities to Logical Formulas", in Hintikka and Suppes (eds.) 1966, 219–264.
Scott D., 1979, "A Note on Distributive Normal Forms", in Saarinen et al. (eds.) 1979, 75-90.
Sellars W., 1947, "Epistemology and the New Way of Words", *The Journal of Philosophy* 44, 645-60.
Sellars W., 1948/1949, "Realism and the New Way of Words", *Philosophy and Phenomenological Research* 8. Reprinted in Feigl and Sellars (eds.) 1949, 424-56.
Sellars W., 1964, "Induction as Vindication", *Philosophy of Science* 31, 197–231.
Shimony A., 1955a, "Coherence and the Axioms of Confirmation", *The Journal of Symbolic Logic* 20, 1-28.
Shimony A., 1955b, "Review of J. Kemeny, A Contribution to Inductive Logic", *The Journal of Symbolic Logic* 20, 189.

Shimony A., 1975, "Carnap on Entropy", in Hintikka (ed.) 1975b, 381-95.

Shimony A., 1988, "An Adamite Derivation of the Principles of the Calculus of Probability", in Fetzer (ed.) 1988, 79-89.

Shimony A., 1992, "On Carnap: Reflections of a Metaphysical Student", *Synthese* 93, 261-274.

Shoemaker S. S., 1980, "Properties, Causation, and Projectibility", in Cohen and Hesse (eds.) 1980, 291-312.

Siegel H., 1980, "Justification, Discovery and the Naturalizing of Epistemology", *Philosophy of Science* 47, 297–321.

Sintonen M., 1997a, "From the Science of Logic to the Logic of Science", in Sintonen (ed.) 1997b, 1-2.

Sintonen M. (ed.), 1997b, *Knowledge and Inquiry: Essays on Jaakko Hintikka's Epistemology and Philosophy of Science*, Amsterdam: Rodopi.

Skyrms B., 1966a, *Choice and Chance: An Introduction to Inductive Logic*, Encino: Dickenson.

Skyrms B., 1966b, "Nomological Necessity and the Paradoxes of Confirmation", *Philosophy of Science* 33, 230-249.

Skyrms B., 1972, "The Goodman Paradox and the New Riddle of Induction", in Feigl et al. (eds.) 1972, 480-90.

Skyrms B., 1983, "Three Ways to Give A Probability Assignment A Memory", in Earman (ed.) 1983, 157-162.

Skyrms B., 1987a, "Dynamic Coherence", in: I. B. MacNeill, and G. Umphrey (eds.), *Foundations of Statistical Inference,* Dordrecht: Reidel, 233-243.

Skyrms B., 1987b, "Dynamic Coherence and Probability Kinematics", *Philosophy of Science* 54, 1-20.

Skyrms B., 1991, "Carnapian Induction Logic for Markov Chains", *Erkenntnis* 35, 439-460.

Skyrms B., 1992, "Coherence, Probability and Induction in Rationality", in E. Villanueva (ed.) 1992, *Epistemology*, Atascadero: Ridgeview, 215-226.

Skyrms B., 1993a, "A Mistake in Dynamic Coherence Arguments?", *Philosophy of Science* 60, 320-328.

Skyrms B., 1993b, "Analogy by Similarity in Hyper-Carnapian Inductive Logic", in J. Earman, A. I. Janis, and G. J. Massey (eds.) 1993, *Philosophical Problems of the Internal and External Worlds: Essays in the Philosophy of Adolf Grünbaum*, Pittsburgh: University of Pittsburgh Press, 273-282.

Skyrms B., 1993c, "Carnapian Inductive Logic for a Value Continuum", in H. Wettstein (ed.) 1993, *The Philosophy of Science,* Midwest Studies in Philosophy, Vol. XVIII, South Bend: University of Notre Dame Press, 78-89.

Skyrms B., 1993d, "Discussion: A Mistake in Dynamic Coherence Arguments?", *Philosophy of Science* 60, 320-328.

Skyrms B., 1993e, "Probability, Theories of", in Dancy and Sosa (eds.) 1993, 374-378.

Skyrms B., 1994, "Bayesian Projectibility", in Stalker (ed.) 1994, 241-262.

Skyrms B., 1996a, "Carnapian Inductive Logic and Bayesian Statistics", in: T. S. Ferguson, L. S. Shapley, and J. B. MacQueen (eds.) 1996, *Statistics, Probability and Game Theory: Papers in Honor of David Blackwell*, Hayward: Institute of Mathematical Statistics, 321-36.

Skyrms B., 1996b, "The Structure of Radical Probabilism", *Erkenntnis* 45, 285-297.

Smokler H., 1967, "Language Dependencies and Systems of Confirmation", *The Journal of Philosophy* 64, 686.

Smokler H., 1977, "Semantical Questions in Carnap's Inductive Logic", *British Journal for the Philosophy of Science* 28, 129-135.
Sober E., 1978, "Psychologism", *Journal for the Theory of Social Behavior* 8, 165-91.
Sosa E., 1991, *Knowledge in Perspective*, Cambridge: Cambridge University Press.
Sosa E., 1998, "Foundationalism", in Craig (ed.) 1998, Vol. III, 718-23.
Spohn W., 1981, "Analogy and Inductive Logic: A Note on Niiniluoto", *Erkenntnis* 16, 35-52.
Spohn W. (ed.), 1991a, *Erkenntnis Orientated: A Centennial Volume for Rudolf Carnap and Hans Reichenbach*, Dordrecht: Kluwer.
Spohn W., 1991b, "Preface", in Spohn (ed.) 1991a, 1-4.
Stalker D. (ed.), 1994, *Grue! The New Riddle of Induction*, La Salle: Open Court.
Stanovich K. E., and West F. R., 2001, "Individual Differences in Reasoning: Implications for the Rationality Debate?", *Behavioral and Brain Sciences* 23, 645-65.
Stegmüller W., 1973, *Carnap II: Normative Theorie des Induktiven Räsonierens*, Berlin: Springer.
Stegmüller W., 1975, "Homage to Carnap", in Hintikka (ed.) 1975b, LIV-LVIII.
Stich S. P., and Nisbett E., 1980/1997, "Justification and Psychology of Human Reasoning", *Philosophy of Science* 47, 188-202. Reprinted in Elgin (ed.) 1997, 274-288.
Stroud B., 1981, "The Significance of Naturalized Epistemology", in P. A. French, T. E. Uehling, and H. K. Wettstein (eds.) 1981, *Midwest Studies in Philosophy VI*, Minneapolis: University of Minnesota, 455-71.
Sturgeon S., 1997, "Reliabilism", in Wójcicki (ed.) 1997, 127-48.
Suppes P., 1966a, "A Bayesian Approach to the Paradoxes of Confirmation", in Hintikka and Suppes (eds.) 1966, 198–207.
Suppes P., 1966b, "Probabilistic Inference and the Concept of Total Evidence", in Hintikka and Suppes (eds.) 1966, 49–65.
Swain M. (ed.), 1970, *Induction, Acceptance, and Rational Belief*, Dordrecht: Reidel.
Swain M., 1981, *Reasons and Knowledge*, Ithaca: Cornell University Press.
Swinburne R., 1971, "The Probability of Particular Events", *Philosophy of Science* 38, 327-43.
Swinburne R. G., 1980, "Properties, Causation/projectibility: Reply to Shoemaker", in Cohen and Hesse (eds.) 1980, 313-320.
Swinburne R. G., 1997, "Justification and Probability", in Wójcicki (ed.) 1997, 149-67.
Tarski A., 1933/1956b, "The Concept of Truth in Formalized Languages". Translated and reprinted in Tarski 1956, 152-278.
Tarski A., 1944/1949, "The Semantic Conception of Truth and the Foundations of Semantics", *Philosophy and Phenomenological Research* 4. Reprinted in Feigl and Sellars (eds.) 1949, 52-84.
Tarski A., 1956, *Logic, Semantics, Metamathematics: Papers From 1923 to 1938*. Translated by J. H. Woodger, Oxford: Clarendon.
Thiel C., 1982, "From Leibniz to Frege: Mathematical Logic Between 1679 and 1879", in Cohen (ed.) 1982, 755-770.
Thom R., 1982, "The Role of Mathematics in Present-Day Science", in Cohen (ed.) 1982, 3-13.
Toulmin S., 1958, *The Uses of Argument*, Cambridge: Cambridge University Press.
Tuomela R., 1966, "Inductive Generalization in an Ordered Universe", in Hintikka and Suppes (eds.) 1966, 155–174.
Tuomela R., 1992, "Group Beliefs", *Synthese* 91, 285-318.

Tversky A., and Kahneman D., 1977, "Causal Thinking in Judgment under Uncertainty", in Hintikka and Butts (eds.) 1977, 167-90.

Tversky A., and Kahneman D., 1983, "Extensional Versus Intuitive Reasoning: The Conjunction Fallacy in Probability Judgment", *Psychological Review* 90, 293–315.

Uchii S., 1972, "Inductive Logic With Causal Modalities: A Probabilistic Approach", *Philosophy of Science* 39, 162-178.

Uchii S., 1973, "Inductive Logic With Causal Modalities: A Deterministic Approach", *Synthese* 26, 264-303.

Uebel T., 1996, "Anti-Foundationalism and the Vienna Circle's Revolution in Philosophy", *British Journal for the Philosophy of Science* 47, 415-40.

Venn J., 1888/1972, *The Principles of Empirical or Inductive Logic*, New York: B. Franklin.

Vickers J. M., 1965, "Some Remarks on Coherence and Subjective Probability", *Philosophy of Science* 32, 32-38.

Vickers J. M., 1989, "An Agenda for Subjectivism", *Erkenntnis* 31, 397-416.

Voishvillo E. K., 1982, "Semantics of Generalized State Descriptions", in Cohen (ed.) 1982, 315-23.

von Plato J., 1981a, "Finite Partial Exchangeability", *Statistics and Probability Letters* 11, 99-102.

von Plato J., 1981b, "On Partial Exchangeability As A Generalization of Symmetry Principles", *Erkenntnis* 16, 53-59.

Wagner C. G., 1997, "Old Evidence and New Explanation", *Philosophy of Science* 64, 677-91.

Walk K., 1966, "Simplicity, Entropy, and Inductive Logic", in Hintikka and Suppes (eds.) 1966, 66–80.

Walker E. R., 1947, "Verification and Probability", *The Journal of Philosophy* 44, 97-104.

Watt D. E., 1989, "Not Very Likely: A Reply to Ramsey", *British Journal for the Philosophy of Science* 40, 223-227.

Weir A., 1995, "Gruesome Perceptual Spaces", *Analysis* 55, 27-36.

Williams D., 1945, "On the Derivation of Probabilities from Frequencies", *Philosophy and Phenomenological Research* 5, 449-484.

Williams D. C., 1952, "Professor Carnap's Philosophy of Probability", *Philosophy and Phenomenological Research* 13, 103-121.

Williamson T., 1997, "Knowledge as Evidence", *Mind* 106, 717-41.

Williamson T., 1998, "Conditioning on Knowledge", *British Journal for the Philosophy of Science* 49, 89-121.

Woleński J., 1997, "Logika i indukcja", in *Szkoła Lwowsko-Warszawska w polemikach*, Warszawa: Scholar, 35-43.

Wójcicki R., 1966/1977, "Semantical Criteria of Empirical Meaningfulness", *Studia Logica* 19. Reprinted in Przełęcki and Wójcicki (eds.) 1977, 647-80.

Wójcicki R., 1976, "Some Problems of Formal Methodology", in Przełęcki et al. (eds.) 1976, 9-18.

Wójcicki R., 1995/1996a, "Theories, Theoretical Models, Truth (I)", *Foundations of Science* 1, 337-406.

Wójcicki R., 1995/1996b, "Theories, Theoretical Models, Truth, Part II: Tarski's Theory of Truth and Its Relevance for the Theory of Science", *Foundations of Science* 1, 471-516.

Wójcicki R. (ed.), 1997a, *Epistemic Justification*, Warszawa: IFiS PAN.

Wójcicki R., 1997b, Main Topics in the Theory of Knowledge: A Historical Introduction, manuscript.

Worrall J., 1989, "Structural Realism: the Best of Both Words?", *Dialectica* 43, 99-124.

Wright G. H. von, 1966, "The Paradoxes of Confirmation", in Hintikka and Suppes (eds.) 1966, 208–218.

Wright G. H. von, 1990, "Wittgenstein and the Twentieth Century", in Haaparanta et al. (eds.) 1990, 47-67.

Zabell S. L., 1982, "W. E. Johnson's "Sufficientness" Postulate", *Annals of Statistics* 10, 1091-99.

Zabell S. L., 1989, "The Rule of Succession", *Erkenntnis* 31, 283-321.

Zabell S. L., 1992, "Predicting the Unpredictable", *Synthese* 90, 205-32.

Zabell S. L., 1995, "Characterizing Markov Exchangeable Sequences", *Journal of Theoretical Probability* 8, 175-78.

Zabell S. L., 1996, "The Continuum of Inductive Methods Revisited", in J. Earman, and J. Norton (eds.), *The Cosmos of Science, Pittsburgh-Konstanz Series in the Philosophy and History of Science*, Pittsburgh\Konstanz: University of Pittsburgh Press\Universitätsverlag Konstanz, 351-85.

Zabell S. L., 1997, "Confirming Universal Generalizations", *Erkenntnis* 45, 267-83.

Index

TRENDS IN LOGIC

1. G. Schurz: *The Is-Ought Problem.* An Investigation in Philosophical Logic. 1997 ISBN 0-7923-4410-3
2. E. Ejerhed and S. Lindström (eds.): *Logic, Action and Cognition.* Essays in Philosophical Logic. 1997 ISBN 0-7923-4560-6
3. H. Wansing: *Displaying Modal Logic.* 1998 ISBN 0-7923-5205-X
4. P. Hájek: *Metamathematics of Fuzzy Logic.* 1998 ISBN 0-7923-5238-6
5. H.J. Ohlbach and U. Reyle (eds.): *Logic, Language and Reasoning.* Essays in Honour of Dov Gabbay. 1999 ISBN 0-7923-5687-X
6. K. Došen: *Cut Elimination in Categories.* 2000 ISBN 0-7923-5720-5
7. R.L.O. Cignoli, I.M.L. D'Ottaviano and D. Mundici: *Algebraic Foundations of many-valued Reasoning.* 2000 ISBN 0-7923-6009-5
8. E.P. Klement, R. Mesiar and E. Pap: *Triangular Norms.* 2000 ISBN 0-7923-6416-3
9. V.F. Hendricks: *The Convergence of Scientific Knowledge.* A View From the Limit. 2001 ISBN 0-7923-6929-7
10. J. Czelakowski: *Protoalgebraic Logics.* 2001 ISBN 0-7923-6940-8
11. G. Gerla: *Fuzzy Logic.* Mathematical Tools for Approximate Reasoning. 2001 ISBN 0-7923-6941-6
12. M. Fitting: *Types, Tableaus, and Gödel's God.* 2002 ISBN 1-4020-0604-7
13. F. Paoli: *Substructural Logics: A Primer.* 2002 ISBN 1-4020-0605-5
14. S. Ghilardi and M. Zawadowki: *Sheaves, Games, and Model Completions.* A Categorical Approach to Nonclassical Propositional Logics. 2002 ISBN 1-4020-0660-8
15. G. Coletti and R. Scozzafava: *Probabilistic Logic in a Coherent Setting.* 2002 ISBN 1-4020-0917-8; Pb: 1-4020-0970-4
16. P. Kawalec: *Structural Reliabilism.* Inductive Logic as a Theory of Justification. 2002 ISBN 1-4020-1013-3

KLUWER ACADEMIC PUBLISHERS – DORDRECHT / BOSTON / LONDON

MIX
Papier aus verantwortungsvollen Quellen
Paper from responsible sources
FSC® C105338

If you have any concerns about our products,
you can contact us on
ProductSafety@springernature.com

In case Publisher is established outside the EU,
the EU authorized representative is:
Springer Nature Customer Service Center GmbH
Europaplatz 3, 69115 Heidelberg, Germany

Printed by Libri Plureos GmbH
in Hamburg, Germany